FADING DISPERSIVE
COMMUNICATION CHANNELS

Fading Dispersive Communication Channels

ROBERT S. KENNEDY

Department of Electrical Engineering
Massachusetts Institute of Technology

WILEY-INTERSCIENCE a division of John Wiley & Sons
New York · London · Sydney · Toronto

To LETTY
Carole, Nancy, and Nina

Preface

This book can be characterized in several ways. At one extreme it represents a straightforward application of the results of detection and information theory to an important class of communication channel models. At another extreme it demonstrates that some real channels can be made to yield better performance than has traditionally been thought possible. A third characterization is that it illustrates how, with judicious approximation, the concepts of information and detection theory can lead to a better understanding of some real communication channels.

The reader's categorization of the book will be strongly influenced by his background. To those familiar with random processes, communication theory, and information theory as presented in Davenport and Root's *Random Signals and Noise*, Wozencraft and Jacobs' *Principles of Communication Engineering*, and Gallager's *Information Theory and Reliable Communication*, it will indeed fall in the first category. To those who characterize communication systems only by input signal-to-noise ratio, it may fall, obscurely, into the second category. I hope that those familiar with Sakrison's *Communication Theory: Transmission of Waveforms and Digital Information* or with the first four chapters of Wozencraft and Jacobs' *Principles of Communication Engineering* will find it digestible.

Most of the results presented in Chapters 5 and 6 were developed in 1963 in conjunction with the Lincoln Laboratory West Ford Project and have not been published previously. The decision to present them here rather than publish them immediately as a series of specialized journal articles was guided by a desire to make them more useful to practicing communication engineers. To this end the first four chapters attempt to relate the gross characterization of real fading dispersive channels to the mathematical model on which the later chapters are based.

In a broad sense this book can be viewed as an extension of the work presented by Robert Price, George Turin, John Pierce, and Ira Jacobs. In a more personal sense it reflects the perspective that I have received from Professors Wozencraft and Siebert and from my involvement with Lincoln Laboratory and the Research Laboratory of Electronics. I benefited from numerous discussions with Estil Hoversten and Kenneth Jordan during the formulation and preparation of the manuscript. I also appreciated the technical comments and personal encouragement of Robert Price, Irwin Lebow, Robert Gallager, Phillip Bello, John Moldon, John Richters, Sherman Karp, Thomas Seay and Seppo Halme.

The manuscript was typed, and retyped, by Mrs. Eleanor Montesano, Mrs. Betty Hill, and Miss Marilyn Pierce.

Robert S. Kennedy

Cambridge, Massachusetts
September, 1968

Contents

FADING DISPERSIVE
COMMUNICATION CHANNELS

Chapter 1

Introduction

The engineering importance of communication media that exhibit both fading and dispersive characteristics has increased markedly in recent years. Several factors have given rise to this increase. On the one hand, theoretical and technological advances have permitted a more efficient utilization of fading ionospheric radio links. Also, new communication media, such as the lunar surface and plasmas, have been employed. In some instances, such as Project West Ford, they have been created.

Detection theory has played an important role in the development of systems employing such channels. The relative merits of various diversity-combining schemes have been evaluated [1–26], optimum receiver structures for digital signaling have been determined [27–40], and the performance of binary signaling systems has been thoroughly investigated [41–59]. However, until recently, larger signaling alphabets have received less attention [60–64]. One of our primary objectives in this monograph is to determine the performance characteristics of large alphabet systems. A secondary objective is to illustrate how the concepts and techniques of modern communication and information theory may be used to advantage by the practicing communication engineer. Thus, in a broad sense, the content of this monograph may be regarded as a particular application of the general concepts and techniques employed.

The channels we consider are characterized, mathematically, by the statement: conditioned on the transmitted waveform, the received waveform is a sample function of a zero-mean Gaussian random process. This characterization provides an adequate base for the development of the communication theory results we require, but these results acquire more substance and can be utilized more effectively when the mathematical characterization of the channel is related to its physical and observable properties. Therefore we shall establish those relations that are relevant to the mathematical model we employ. Comprehensive discussions of the general physical properties of many important channels are available elsewhere [65–72].

1

1.1 THE COMMUNICATION SYSTEM

The general structure of the communication system to be considered is depicted in Fig. 1.1. As illustrated, the input to the transmitting terminal and the output of the receiving terminal consist of sequences of binary messages, or bits, which occur at a rate of R bits/sec. Roughly stated, we characterize the system by an error probability $P(\varepsilon)$ and determine the dependence of this probability upon the rate R and upon the characteristics of the transmitter, the channel, and the receiver.

The communication media, or channels, we consider may be specified by a scattering function $\sigma(r, f)$. This function describes the distribution of reflection cross section in time delay and in Doppler shift; that is, we envisage the medium as though it were composed of a large collection of point scatterers, for examples, dipoles or differential elements of the lunar surface. Each of these point scatterers is characterized, in part, by a time delay, a Doppler shift, and an average reflecting cross section. The value of $\sigma(r, f)\, dr\, df$ is then the fraction of the total average cross section contributed by the scatterers whose range delay is between r and $r + dr$ and whose Doppler shift is between f and $f + df$.

As shown in Fig. 1.1, we envisage the transmitter and the receiver as each composed of two subunits. The first, and more familiar, units are the modulator and demodulator. For our purposes the modulator is described by the

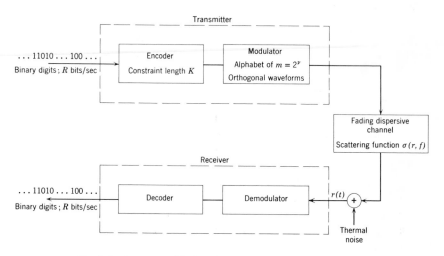

Fig. 1.1 Structure of fading dispersive communication system.

number m and the character of the waveforms employed. One of our objectives is to determine those characteristics of the waveforms which most influence the system's performance. The principal constraint employed is that the waveforms be orthogonal in the sense defined in Chapter 4.

When K equals v, the system performance improves as v is increased, but the complexity of the demodulator eventually becomes intolerable; also the required system bandwidth becomes immense. These problems can be circumvented, to some degree, by increasing K with v fixed so as to introduce constraints between successive transmissions of waveforms selected from a smaller alphabet. These constraints are allowed for by including the encoder and decoder in Fig. 1.1. The principal supposition is that the encoder and modulator are designed so that successively transmitted waveforms are independently affected by the channel; that is, there is no intersymbol interference or memory. Although we do not analyze coded systems, we develop them to the point at which the existing techniques and results of coding theory can be applied. As an illustration of these results, the performance that can be achieved with block coding is included in Chapter 5.

The channel model employed describes many existing fading dispersive communication channels, but the constraints imposed upon the encoder and modulator do limit the utility of the results. Broadly stated, this monograph provides a rather comprehensive treatment of the performance levels that can be attained when the desired information rate is appreciably less than the available system bandwidth and is not too much greater than the reciprocal of the channel time dispersion or "multipath spread"; for example, it is directly applicable to many tropospheric scatter channels. On the other hand, the analytical results of the later chapters are not directly relevant to high-rate high-frequency (HF) systems for which the bandwidth constraints are quite severe. Nor do they apply to some important problems, such as the atmospheric optical channel, for which our model is inappropriate. However the first four chapters provide a background that is as applicable to HF and atmospheric systems as it is to tropospheric systems. Moreover the analysis of Chapters 5 through 7 provides an introduction to the analysis of narrower-band systems and other models [73–75].

The character of the results we obtain is illustrated in Fig. 1.2 for the situation in which K equals v. As noted in the figure, the error probability is given approximately by an expression of the form 2^{-KE_b} where the function E_b is plotted versus β, the ratio of average received energy per information bit to noise power density. The curve labeled "thermal noise" specifies the value of E_b under the assumption that the communication channel is neither fading nor dispersive; the remaining curve pertains to the fading dispersive channels that are considered in this monograph.

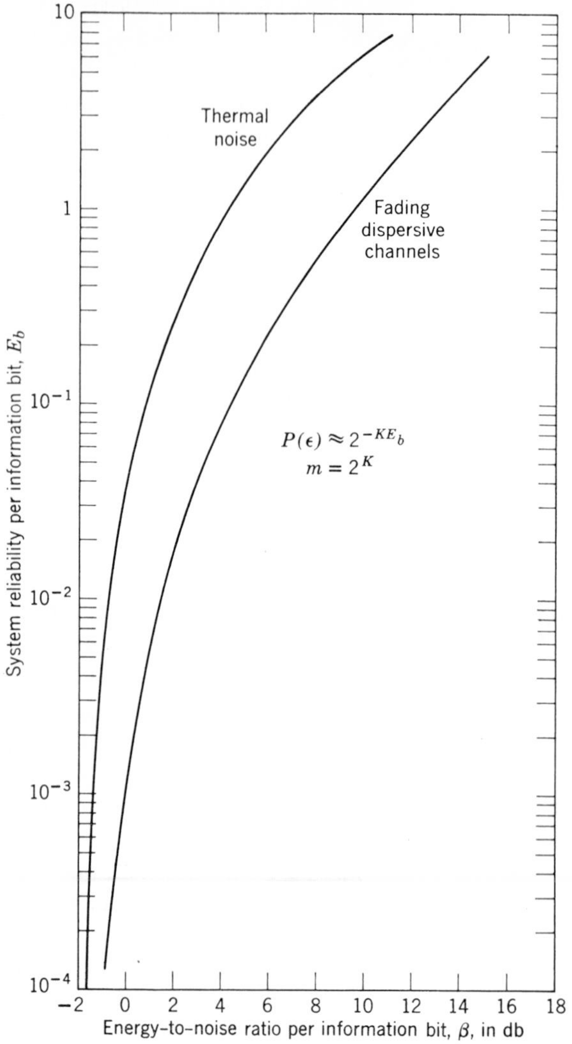

10
1
10^{-1}
10^{-2}
10^{-3}
10^{-4}
System reliability per information bit, E_b
Thermal noise
Fading dispersive channels
$P(\epsilon) \approx 2^{-KE_b}$
$m = 2^K$
−2 0 2 4 6 8 10 12 14 16 18
Energy-to-noise ratio per information bit, β, in db

Fig. 1.2 Optimized error probability for uncoded orthogonal signals.

1.2 SUMMARY OF CHAPTERS

Chapter 2 is devoted to a more detailed description of the communication media we consider. In Chapter 3 is presented a discussion of the gross observable characteristics of fading dispersive channels; it is included to provide a link between the mathematical model employed and the observable phenomena of the real world. It also provides some crude estimates of important quantities that are laborious to determine precisely.

In Chapter 4 the constraints imposed upon the communication system portrayed in Fig. 1.1 are discussed in greater detail. Other system considerations, such as intersymbol interference and bandwidth requirements, are also considered there.

Chapter 5 is devoted to a discussion of the ultimate performance limitations of fading dispersive systems when v equals K. The results presented relate the minimum attainable error probability to R, the information rate; K, the constraint length; and β, the ratio of average received energy per information bit to noise power density. The distinguishing characteristics of systems that achieve the minimum error probability are also discussed. Since these characteristics involve the properties of both the channel and the transmitted waveforms, the results of Chapter 5 do not indicate the performance that can, in fact, be achieved for a given scattering function.

In Chapter 6 we attack the problem of determining the performance that can be achieved for a given fading dispersive channel. The problem is a very complex one and our results are not complete, but it appears that the performance levels described in Chapter 5 can be achieved in the engineering sense for a very large variety of scattering functions. This is true for both overspread and underspread channels. Moreover the design of "good" systems does not depend upon the detailed structure of the scattering functions. In particular, the maximum information rate at which reliable communication can be sustained is independent of the scattering function and equals the capacity of a nonfading nondispersive channel.

Chapters 5 and 6 are predicated upon the assumption that the best possible demodulator is employed. This is unrealistic because the optimum demodulator is, in general, both difficult to determine and impractical to build. Therefore in Chapter 7 we consider the degradation in performance that results when other, more appealing, demodulators are employed. The results indicate that the loss need not be substantial.

REFERENCES

[1] G. F. Montgomery, "Message Error in Diversity Frequency-Shift Reception." *Proc IRE*, 1184–1187, July 1954.
[2] L. R. Kahn, "Ratio Squarer." *Proc. IRE*, 1704, November 1954.
[3] D. G. Brennan, "On the Maximum Signal-to-Noise Ratio Realizable from Several Noisy Signals." *Proc. IRE*, 1530, October 1955.
[4] H. Staras, "The Statistics of Combiner Diversity." *Proc. IRE*, 1057–1058, August 1956.
[5] J. N. Pierce, "Theoretical Diversity Improvement in Frequency-Shift Keying." *Proc IRE*, 903–910, May 1958.
[6] D. G. Brennan, "Linear Diversity Combining Techniques." *Proc. IRE*, 1075–1102, June 1959.

[7] J. N. Pierce and S. Stein, "Multiple Diversity with Nonindependent Fading." *Proc. IRE*, 89–104, January 1960.

[8] J. N. Pierce, "Theoretical Limitations on Frequency and Time Diversity for Fading Binary Transmissions." *IRE Trans. Commun. Systems*, 186–189, June 1961.

[9] G. L. Turin, "On Optimal Diversity Reception." *IRE Trans. Inform. Theory*, 154–166, July 1961.

[10] G. L. Turin, "On Optimal Diversity Reception, II." *IRE Trans. Commun. Systems*, 22–31, March 1962.

[11] P. A. Bello and B. D. Nelin, "Predetection Diversity Combining with Selectively Fading Channels." *IRE Trans. Commun. Systems*, 32–42, March 1962.

[12] P. M. Hahn, "Theoretical Diversity Improvement in Multiple Frequency Shift Keying." *IRE Trans. Commun. Systems*, 177–184, June 1962.

[13] R. Price, "Error Probabilities for Adaptive Multichannel Reception of Binary Signals." *IEEE Trans. Inform. Theory*, 305–316, September 1962.

[14] J. N. Pierce, "Approximate Error Probabilities for Optimal Diversity Combining." *IEEE Trans. Commun. Systems*, 352–354, September 1963.

[15] W. C. Lindsey, "Asymptotic Performance Characteristics for the Adaptive Coherent Multireceiver and Noncoherent Multireceiver Operating Through the Rician Fading Multichannel." *IEEE Trans. Commun. Electron*, pp. 67–73, January 1964.

[16] W. C. Lindsey, "Comparison of Nonlinear and Linear Multireceiver Detection Systems," *IEEE Trans. Space Electron. Telemetry*, 10–14, March 1964.

[17] M. Nesenbergs, "Binary Error Probability Due to an Adaptable Fading Model." *IEEE Trans. Commun. Systems*, 64–73, March 1964.

[18] W. C. Lindsey, "Error Probabilities for Rician Fading Multichannel Reception of Binary and N-ary Signals." *IEEE Trans. Inform. Theory*, 339–350, October 1964.

[19] P. R. Reed and J. N. Pierce, "Comparison of Square-Law and Majority-Count Diversity Combiners." *IEEE Trans. Commun. Technology*, 217, December 1964.

[20] W. C. Lindsey, "Error Probability for Incoherent Diversity Reception." *IEEE Trans. Inform. Theory*, 491–499, October 1965.

[21] J. N. Pierce, "Plurality-Count Diversity Combining for Fading M-ary Transmissions." *IEEE Trans. Commun. Technology*, 529–532, August 1966.

[22] W. C. Lindsey, "Error Probabilities for Partially Coherent Diversity Reception." *IEEE Trans. Commun. Technology*, 620–625, October 1966.

[23] I. M. Jacobs, "Probability-of-Error Bounds for Binary Transmission on the Slowly Fading Rician Channel." *IEEE Trans. Inform. Theory*, 431–441, October 1966.

[24] R. T. Aitken, "Communication Over the Discrete-Path Fading Channel." *IEEE Trans. Inform. Theory*, 346–348, April, 1967.

[25] N. T. Gaarder, "Maximal-Ratio Diversity Combiners." *IEEE Trans. Commun. Technology*, 790–796, December 1967.

[26] J. G. Proakis, "On the Probability of Error for Multichannel Reception of Binary Signals." *IEEE Trans. Commun. Technology*, 68–70, February 1968.

[27] R. Price, "The Detection of Signals Perturbed by Scatter and Noise." *IRE Trans. Inform. Theory*, 163–170, September 1954.

[28] R. Price, "Optimum Detection of Random Signals in Noise, with Application to Scatter-Multipath Communication, I." *IRE Trans. Inform. Theory*, 125–135, December 1956.

[29] G. L. Turin, "Communication through Noisy, Random-Multipath Channels." *IRE Convention*, 154–166, 1956.

[30] D. Middleton, "On the Detection of Stochastic Signals in Additive Normal Noise—Part I." *IRE Trans. Inform. Theory*, 86–121, June 1957.

[31] R. Price and P. E. Green, "Communication Technique for Multipath Channels." *Proc. IRE*, 555–570, March 1958.

[32] D. Slepian, "Some Comments on the Detection of Gaussian Signals in Gaussian Noise." *IRE Trans. Inform. Theory*, 65–68, June 1958.

[33] D. Middleton, "On New Classes of Matched Filters and Generalizations of the Matched Filter Concept." *IRE Trans. Inform. Theory*, 349–360, June 1960.

[34] T. Kailath, "Correlation Detection of Signals Perturbed by a Random Channel." *IRE Trans. Inform. Theory*, 361–366, June 1960.

[35] C. Cherry, Ed., *Information Theory*, Fourth London Symposium. Thomas Kailath, "Optimum Receivers for Randomly Varying Channels," Washington, D.C.: Butterworths, 1961, p. 109.

[36] S. M. Sussman, "A Matched Filter Communication System for Multipath Channels." *IRE Trans. Inform. Theory*, 367–373, June 1960.

[37] P. A. Bello, "Some Results on the Problem of Discriminating Between Two Gaussian Processes." *IRE Trans. Information Theory*, 224–233, October 1961.

[38] D. R. Bitzer, D. A. Chesler, R. Ivers, and S. Stein, "A Rake System for Tropospheric Scatter." *IEEE Trans. Commun. Technology*, 499–506, August 1966.

[39] R. W. Chang, "On Receiver Structures for Channels Having Memory." *IEEE Trans. Inform. Theory*, 463–468, October 1966.

[40] T. T. Kadota and L. A. Shepp, "On the Best Finite Set of Linear Observables for Discriminating Two Gaussian Signals." *IEEE Trans. Inform. Theory*, 278–284, April 1967.

[41] G. L. Turin, "Error Probabilities for Binary Symmetric Ideal Reception through Nonselective Slow Fading and Noise." *Proc. IRE*, 1603–1619, September 1958.

[42] G. L. Turin, "Some Computations of Error Rates for Selectively Fading Multipath Channels." N.E.C., 1959, pp. 431–440.

[43] B. B. Barrow, "Error Probabilities for Telegraph Signals Transmitted on a Fading FM Carrier." *Proc. IRE*, 1613–1629, September 1960.

[44] P. A. Bello, "The Influence of Fading Spectrum on the Binary Error Probabilities of Incoherent and Differentially Coherent Matched Filter Receivers." *IRE Trans. Commun. Systems*, 160–168, June 1962.

[45] P. A. Bello and B. D. Nelin, "The Effect of Frequency Selective Fading on the Binary Error Probabilities of Incoherent and Differentially Coherent Matched Filter Receivers." *IEEE Trans. Commun. Systems*, 170–186, June 1963.

[46] J. C. Hancock, "Optimum Performance of Self-Adaptive Systems Operating Through a Rayleigh-Fading Medium." *IEEE Trans. Commun. Systems*, 443–453, December 1963.

[47] J. N. Pierce, "Error Probabilities for a Certain Spread Channel." *IEEE Trans. Commun. Systems*, 120–122, March 1964.

[48] R. S. Kennedy and I. L. Lebow, "Signal Design for Dispersive Channels." *IEEE Spectrum*, 231–237, March 1964.

[49] P. A. Bello and B. D. Nelin, "The Effect of Frequency Selective Fading on Intermodulation Distortion and Subcarrier Phase Stability in Frequency Modulation Systems." *IEEE Trans. Commun. Systems*, 87–101, March 1964.

[50] P. A. Bello and B. D. Nelin, "Optimization of Subchannel Data Rate in FDM-SSB Transmission over Selectively Fading Media." *IEEE Trans. Commun. Systems*, 46–53, March 1964.

[51] J. G. Proakis, P. R. Drouilhet, Jr., and R. Price, "Performance of Coherent Detection Systems Using Decision-Directed Channel Measurement." *IEEE Trans. Commun. Systems*, 54–63, March 1964.

[52] W. F. Walker, "The Error Performance of A Class of Binary Communications Systems in Fading and Noise." *IEEE Trans. Commun. Systems*, 28–45, March 1964.

[53] G. D. Hingorani and J. C. Hancock, "A Transmitted Reference System for Communication in Random or Unknown Channels." *IEEE Trans. Commun. Technology*, 293–301, September 1965.

[54] R. Esposito, D. Middleton, and J. A. Mullen, "Advantage of Amplitude and Phase Adaptivity in the Detection of Signals Subject to Slow Rayleigh Fading." *IEEE Trans. Inform. Theory*, 473–481, October 1965.

[53] L. Filippov and V. Smoljaninov, "Optimum Recognition of Binary Signals Passed Through Channels with Random Parameters. *IEEE Trans. Inform. Theory*, 244–248, April 1966.

[56] P. A. Bello, "Binary Error Probabilities Over Selectively Fading Channels Containing Specular Components." *IEEE Trans. Commun. Technology*, 400–406, August 1966.

[57] N. J. Bershad, "Optimum Binary FSK for Transmitted Reference Systems Over Rayleigh Fading Channels." *IEEE Trans. Commun. Technology*, 784–790, December 1966.

[58] R. Esposito, "Error Probabilities for the Nakagami Channel." *IEEE Trans. Inform. Theory*, 145–148, January 1967.

[59] G. D. Hingorani, "Error Rates for a Class of Binary Receivers." *IEEE Trans. Commun. Technology*, 209–215, April 1967.

[60] I. Jacobs, "The Asymptotic Behavior of Incoherent M-ary Communication Systems." *Proc. IEEE*, 251–252, January 1963.

[61] H. L. Yudkin, "An Error Bound for Gaussian Signals in Gaussian Noise." M.I.T. Research Laboratory of Electronics, Q.P.R. No. 73, 149, April 1964.

[62] R. S. Kennedy, "Performance Limitations of Dispersive Fading Channels," ICMCI, Tokyo, Japan, 1964, Part 3, 61.

[63] J. Ziv, "Probability of Decoding Error for Random Phase and Rayleigh Fading Channels." *IEEE Trans. Inform. Theory*, 53–61, January 1965.

[64] A. J. Viterbi, "Performance of an M-ary Orthogonal Communication System Using Stationary Stochastic Signals." *IEEE Trans. Inform. Theory*, **IT-13**, July 1967.

[65] M. Schwartz, W. R. Bennett, and S. Stein, *Communication Systems and Techniques*. New York: McGraw-Hill, 1966, Chapter 9.

[66] J. V. Evans and T. Hagfors, Eds., *Radar Astronomy*. New York: McGraw-Hill, 1968.

[67] U.S. Department of Commerce, *Ionospheric Radio Propagation*, 1965.

[68] K. Fokestad, *Ionospheric Radio Communications*. New York: Plenum Press, 1968.

[69] "Scatter Issue," *Proc. IEEE*, October 1955.

[70] Meteor-Burst Communication Papers, *Proc. IRE*, December 1957.

[71] J.T.A.C., "Radio Transmission by Ionospheric and Tropospheric Scatter." *Proc. IRE*, 4, January 1960.

[72] "Westford Issue," *Proc. IEEE*, May 1964.

[73] J. S. Richters, "Communication Over Fading Dispersive Channels." M.I.T. Research Laboratory of Electronics, Technical Report No. 464, November 30, 1967.

[74] R. S. Kennedy and E. V. Hoversten, "On the Atmosphere as an Optical Communication Channel." *IEEE Trans. Inform. Theory*, September 1968.

[75] S. Halme, "Efficient Optical Communication through a Turbulent Atmosphere." M.I.T. Research Laboratory of Electronics, Q.P.R., No. 91, October 15, 1968.

Chapter 2

The Channel Model

Fading dispersive channels are usually best described as random linear time-varying filters. In many applications it is reasonable to suppose that the impulse response of the filter is a sample function of a Gaussian random process; an equivalent assumption is that, given the transmitted waveform, the received process is Gaussian. If one accepts this supposition, as we do, the specification of the channel reduces to the specification of the mean and correlation function either of the channels' random impulse response or of the received process conditioned upon the transmitted waveform. Many radio channels are adequately modeled by taking the mean to be zero and the correlation function to be of a special form that is characterized by a scattering function. For simplicity we restrict our attention to such channels.

The correlation function of the received process can be taken as the point of departure for our discussion of the channel model. However, it is helpful to have a picture of a physical channel for which our several suppositions are realistic. To that end we first discuss channels composed of many isolated point scatterers. These channels also provide a physical interpretation of the channel scattering function. Then in Section 2.3 we relate the scattering function to the transmission properties of the channel as measured by the two-frequency correlation function. These relations establish the class of linear time-varying channels to which the model applies and provide more insight into the effect of the channel upon its input. Finally we establish a canonical representation of fading dispersive channels as classical diversity systems and note an equivalence that exists between channels.

Throughout the discussion we denote the channel input and output waveforms by $s(t)$ and $y(t)$, respectively. These waveforms will usually be represented by their *complex envelopes*. Thus, for example, we suppose that

$$s(t) = \text{Re} \left[u(t) \exp j\omega_o t \right], \qquad (2.1)$$

where $u(t)$ is the complex envelope of $s(t)$, Re denotes "the real part" of the indicated quantity, and ω_o denotes the nominal carrier frequency in radians

9

per second. Since we always specify the complex envelopes directly, there should be no difficulties in their use. However, extensive discussions of the relations between waveforms and their complex envelopes are available for the reader desiring further background [1–5].

2.1   THE POINT-SCATTERER DESCRIPTION

As illustrated in Fig. 2.1, we consider a channel in which propagation is established by single scattering from a large number of independent elements, or scatterers. These scatterers might be orbital dipoles or differential elements of either the troposphere or the ionosphere.

We describe each scatterer by its reflection energy cross section ρ_i^2 and by its propagation time delay $T_i(t)$. For convenience we assume that each $T_i(t)$ is a linear function of time; that is,

$$T_i(t) = \tau_i + \dot{\tau}_i t, \tag{2.2}$$

where τ_i, the initial propagation delay, and $\dot{\tau}_i$, the rate of change of delay, are constants independent of time. If the transmitter and receiver are located at the same site this assumption is equivalent to the assumption of a constant radial component of scatterer velocity. More general expressions for the delay can be employed, but it appears that their use does not appreciably

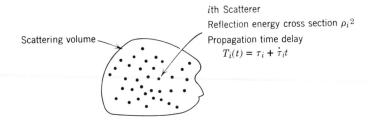

ith Scatterer
Reflection energy cross section ρ_i^2
Propagation time delay
$$T_i(t) = \tau_i + \dot{\tau}_i t$$

Scattering volume

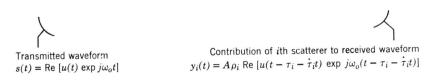

Transmitted waveform
$s(t) = \mathrm{Re}\,[u(t)\,\exp j\omega_o t]$

Contribution of ith scatterer to received waveform
$y_i(t) = A\rho_i\,\mathrm{Re}\,[u(t - \tau_i - \dot{\tau}_i t)\,\exp j\omega_o(t - \tau_i - \dot{\tau}_i t)]$

Fig. 2.1   Geometry of single scattering channel.

affect the applicability of the model to the problems of interest, whereas it does complicate the analysis. Thus we prefer the simple linear expression.

Similar comments apply to the assumption that the energy cross section of a scatterer is constant. We can permit this cross section to be a function of time and thus account for such effects as the time that a given scatterer is "seen" by the antennas. However, it again appears that the additional complication introduced by these generalizations offsets the increased realism they afford. Moreover in the last analysis, the utility of any model is determined by the extent to which the results deduced therefrom coincide with analogous results obtained by experiment. The model employed here appears to give adequate agreement for many problems of interest.

Deterministic Model

Let us first consider the contribution of any single scatterer of the medium to the received waveform. We suppose that the contribution $\tilde{y}_i(t)$ of the ith scatterer is just a delayed and attenuated replica of the transmitted waveform. In particular

$$\tilde{y}_i(t) = A\rho_i \, \text{Re} \, [u(t - \tau_i - \dot{\tau}_i t) \exp j\omega_o(t - \tau_i - \dot{\tau}_i t)], \qquad (2.3)$$

where A is an unimportant constant which is eliminated subsequently.

We now restrict our attention to situations for which the variation of $\dot{\tau}_i t$ in the argument of u may be ignored over the time interval of interest; that is, we suppose that the variation in $\dot{\tau}_i t$ is appreciably less than the reciprocal bandwidth of $u(t)$. This supposition implies that

$$u(t - \tau_i - \dot{\tau}_i t) \approx u(t - \tau_i - \dot{\tau}_i t_o) \qquad (2.4)$$

for any value t_o of t in the interval. In particular we may, without loss of generality, suppose that the time origin is contained in the interval whence

$$u(t - \tau_i - \dot{\tau}_i t) \approx u(t - \tau_i). \qquad (2.5)$$

Stated alternately, we suppose that the time variation in the propagation delay of the ith scatterer does not significantly alter the modulation of its contribution to the received waveform, a supposition that is frequently satisfied in practice [6].

We assume further that the bandwidth of $u(t)$ is much less than the carrier frequency ω_o. This is the conventional "narrow-band" condition encountered in most communication systems. It implies that variations of about π/ω_o in the value of τ_i do not appreciably alter the value of $u(t - \tau_i)$, although they do drastically alter the value of $y_i(t)$ because they change the exponent $\omega_o(t - \tau_i - \dot{\tau}_i t)$ by π radians.

Since we seldom know the value of each τ_i precisely, and since small perturbations in the value are important, it appears reasonable to express each τ_i in the form

$$\tau_i = r_i + \frac{\theta_i}{\omega_o}. \tag{2.6}$$

In this expression we take r_i to be the known, or gross, value of τ_i and account for the perturbations about this value by θ_i, which we take to be a variable whose value lies in the interval $-\pi$ to $+\pi$.

By virtue of (2.5) and (2.6) and the preceding comments, (2.3) can be cast in the form

$$\tilde{y}_i(t) = A \operatorname{Re} \{\rho_i u(t - r_i) \exp j[(\omega_o - \omega_i)t - \omega_o r_i - \theta_i]\}, \tag{2.7a}$$

where

$$\omega_i = \dot{\tau}_i \omega_o. \tag{2.7b}$$

In this expression ω_i is the Doppler shift of the scatterer measured in radians per second, and r_i is as before its mean time, or range, delay measured in seconds.

Equation 2.7 provides an expression for the contribution of the ith scatterer to the received waveform. If secondary scattering is unimportant, as we suppose, the expression for the total received waveform is obtained by summing (2.7) over all the scatterers composing the medium. Specifically

$$y(t) = A \operatorname{Re} \left\{ \sum_i \rho_i u(t - r_i) \exp j[(\omega_o - \omega_i)t - \omega_o r_i - \theta_i] \right\}. \tag{2.8}$$

We have now obtained an expression for the received waveform, but it may involve a very large number of parameters that are either unknown or random in nature. Therefore it is appropriate to resort to a statistical description of the process defined by (2.8); we turn now to that description.

Statistical Model

We first observe that the uncertainty associated with each θ_i corresponds to a delay uncertainty of about half a wavelength. Since the total delay is usually many thousands of wavelengths, the percentage uncertainty is exceedingly small, and there is good reason to assume that if θ_i is to be regarded as a random variable it should be uniformly distributed over the region of uncertainty. Therefore we take each θ_i to be uniformly distributed in the interval $-\pi$ to π. The same argument applied to the difference in delay between scatterers suggests that the θ_i should be presumed to be statistically independent of each other. Thus it only remains to specify the values, or statistical properties, of the scatterer cross sections ρ_i^2.

Since the cross sections ρ_i^2 account for such factors as the aspect of the scatterers, it should be reasonable to treat them as unknown quantities and perhaps as random variables. If we do choose to regard them as random variables, there is little reason to assume that they are statistically dependent upon each other, or that their values are dependent upon the values of the random phase angles θ_i. Thus it is appropriate to regard them as statistically independent random variables, each described by a probability density $p_i(\rho_i)$. We account for nonrandom cross sections in this description by choosing the $p_i(\rho_i)$ to be impulse functions.

The assumptions we have made thus far imply that the mean value $\overline{y(t)}$ of $y(t)$ is zero. They also imply that the correlation function

$$R_y(t, \tau) = \overline{y(t)y(\tau)} \tag{2.9a}$$

of $y(t)$ is given by the expression

$$R_y(t, \tau) = \text{Re}\left[\sum_i \frac{A^2}{2} \overline{\rho_i^2} u(t - r_i)u^*(\tau - r_i) \exp j(\omega_o - \omega_i)(t - \tau)\right], \tag{2.9b}$$

where the asterisk denotes complex conjugation and the bar denotes the ensemble average of the indicated quantity.

The mean and correlation function of $y(t)$ do not, in general, provide a complete statistical description of the received process. There are, however, sound, if not compelling, arguments for supposing that the received process is Gaussian and hence completely described by its mean and correlation function.

Foremost among these is the experimental evidence that suggests that the received process may be Gaussian. Specifically the envelope and phase of the process often exhibit the behavior to be expected from a Gaussian random process [7–14]. Also sometimes there are reasons to believe that the scatterer cross sections ρ_i^2 are chi-squared variates or equivalently that the ρ_i are Rayleigh-distributed random variables. This supposition, in conjunction with our previous assumptions concerning the ρ_i and θ_i, implies that the received process is a Gaussian one.

Alternately, if the number of scatterers is exceedingly large and the cross section of each scatterer is exceedingly small (be it a random value or not), our previous assumptions pertaining to the θ_i ensure that the received process is Gaussian. We render this statement precise in Section 2.5.

The comments of the preceding paragraph, coupled with the difficulties of any other course of action, provide adequate justification for the assumption that the received process is Gaussian; we shall henceforth so assume.

The mathematical description of the received process $y(t)$ is now essentially complete. It is the zero-mean Gaussian random process specified by the correlation function of (2.9). However it is rather awkward to manipulate and interpret this correlation function in its present form, so we prefer instead to express it as an integral involving a density of average energy cross section. Not only is the resulting formulation mathematically expedient, it also provides a more realistic description of fading dispersive channels.

2.2 THE SCATTERING FUNCTION

It is clear that the right member of (2.9b) depends only upon the total average cross section associated with each pair of values of r_i and ω_i; the number of scatterers involved is immaterial. Therefore it is convenient to introduce the average scatterer cross section $\tilde{\sigma}(r, f)$ associated with each such pair, that is,

$$\tilde{\sigma}(r, f) = \sum_i \overline{\rho_i^2}, \qquad (2.10a)$$

where the summation is over all scatterers for which

$$r_i = r \qquad (2.10b)$$

and

$$\omega_i = 2\pi f. \qquad (2.10c)$$

The correlation function $R_y(t, \tau)$ may then be expressed as

$$R_y(t, \tau) = \mathrm{Re}\left[\frac{A^2}{2} \sum u(t - r)u^*(\tau - r)\tilde{\sigma}(r, f) \exp j(\omega_o - \omega)(t - \tau)\right], \qquad (2.11a)$$

where

$$\omega = 2\pi f \qquad (2.11b)$$

and where the summation is over all pairs r and f for which $\tilde{\sigma}(r, f)$ is nonzero.

To avoid questions of mathematical convergence, we assume that the number of scatterers composing the medium is finite. On the other hand, we often may be concerned with situations in which it is impossible to distinguish the contributions to the received waveform of the different scatterers. This occurs when, for any given scatterer with range delay r and Doppler shift f, there are many other scatterers whose range delay differs from r by much less than the reciprocal bandwidth of $u(t)$ and when there are also many scatterers whose Doppler shift differs from f by much less than the reciprocal time duration of $u(t)$.

When this situation occurs the channel behaves as though the point function $\tilde{\sigma}(r, f)$ is replaced by a smooth density and the summation of (2.11) is replaced by an integral. Since attempts to measure the characteristics of fading dispersive channels also yield smooth densities rather than point functions and since we can formally describe a point function by an impulsive density, we henceforth employ the density-function description.†

Specifically we envisage $\tilde{\sigma}(r, f)$ as though it is a density defined for all values of r and f. The cross section associated with values of delay and Doppler in the interval r to $r + dr$ and f to $f + df$ is then $\tilde{\sigma}(r, f) \, dr \, df$, and (2.11a) becomes‡

$$R_y(t, \tau) = \text{Re} \left[\frac{A^2}{2} \iint \tilde{\sigma}(r, f) u(t - r) u^*(\tau - r) \exp j(\omega_o - \omega)(t - \tau) \, dr \, df \right],$$

(2.12)

where $\omega = 2\pi f$.

The function $\tilde{\sigma}(r, f)$ describes both the distribution of average cross section and also the total amount of such cross section. Similarly, $u(t)$ describes both the structure of the transmitted waveform and also its energy level. In each instance the former attribute relates to the general structure of the received process, whereas the latter attribute merely determines the average received energy. We choose now to separate these two attributes of the system in the specification of $R_y(t, \tau)$.

Specifically we henceforth suppose that $u(t)$ has been scaled to unit norm, that is, that

$$\int |u(t)|^2 \, dt = 1.$$

(2.13)

This supposition is not really restrictive since any $u(t)$ can be scaled to satisfy it by redefining A.

We also introduce $\sigma(r, f)$, the normalized density of cross section, or as it is usually called the *channel scattering function*§ [15–19]. It is defined as

$$\sigma(r, f) = \tilde{\sigma}(r, f) \left[\iint \tilde{\sigma}(r, f) \, dr \, df \right]^{-1}.$$

(2.14)

Clearly

$$\iint \sigma(r, f) \, dr \, df = 1.$$

Finally we note that the average received energy E_r is given by each of the expressions

$$E_r = \int \overline{[y(t)]}^2 \, dt = \int R_y(t, t) \, dt = \frac{A^2}{2} \iint \sigma(r, f) \, dr \, df.$$

(2.15)

---

† The problem of scattering function measurement has been approached in several ways [40–45].
‡ The limits of all integrals are $-\infty$ to $+\infty$ unless specifically noted to the contrary.
§ The notation $S(\cdot, \cdot)$, has also been employed [19].

By virtue of (2.13) through (2.15), (2.12) can be expressed as

$$R_y(t, \tau) = \text{Re} \, [E_r R(t, \tau) \exp j\omega_o(t - \tau)], \tag{2.16a}$$

where

$$R(t, \tau) = \iint \sigma(r, f) u(t - r) u^*(\tau - r) \exp j\omega(\tau - t) \, dr \, df. \tag{2.16b}$$

The function $R(t, \tau)$ defined by (2.16b) is, in fact, the complex correlation function of the complex envelope of $y(t)/\sqrt{2E_r}$. That is, if (2.8) is expressed as

$$y(t) = \sqrt{2E_r} \, \text{Re} \, [z(t) \exp j\omega_o t], \tag{2.17a}$$

then

$$\overline{z(t)z^*(\tau)} = R(t, \tau). \tag{2.17b}$$

Equation 2.8 and our assumptions concerning the θ_i also imply that the mean of $z(t)$ is zero and that

$$\overline{z(t)z(\tau)} = 0. \tag{2.17c}$$

We mention these facts because it is occasionally convenient to deal with $z(t)$ in lieu of $y(t)$. Moreover, we usually employ $R(t, \tau)$ in our discussions since it essentially determines $R_y(t, \tau)$.

At this juncture we have developed the mathematical description of the physical scattering channels depicted in Fig. 2.1. They are specified by a scattering function $\sigma(r, f)$, and the received waveform in response to any given transmitted waveform is a sample function of the zero-mean Gaussian random process $y(t)$ whose correlation function is given by (2.16). This model is summarized in Fig. 2.2. We now briefly consider another description of the channel which does not involve the notion of scatterers. This description focuses upon transmission characteristics and tends to distinguish between the roles of the transmitted waveform and of the medium in determining the output correlation function. Either description provides a complete specification of the channels we consider—be they physical scattering channels or not.

2.3   THE TWO-FREQUENCY CORRELATION FUNCTION

The propagation media we have considered constitute linear systems. Consequently if the complex envelope $u(t)$ is expressed as a superposition of elementary complex waveforms, the output waveform is the superposition of the responses to these elementary waveforms. That is, if

$$u(t) = \int u_\alpha(t) \, d\alpha, \tag{2.18a}$$

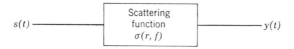

$$s(t) = \text{Re } [u(t) \exp j\omega_0 t] \qquad\qquad y(t) = \sqrt{2E_r} \text{ Re } [z(t) \exp j\omega_0 t]$$

Given $s(t)$, $z(t)$ is a zero-mean complex Gaussian random process with

$$\overline{z(t)z(\tau)} = 0$$

$$\overline{z(t)z^*(\tau)} = R(t, \tau)$$

and

$$R(t, \tau) = \iint \sigma(r, f)u(t - r)u^*(\tau - r) \exp j2\pi f(\tau - t) \, dr \, df$$

$y(t)$ is then a zero-mean Gaussian random process with the correlation function

$$R_y(t, \tau) = \text{Re } [E_r R(t, \tau) \exp j\omega_0(t - \tau)]$$

Fig. 2.2 Summary of point-scatterer model.

then

$$y(t) = \int v_\alpha(t) \, d\alpha, \tag{2.18b}$$

where $v_\alpha(t)$ is the response to $u_\alpha(t)$.

It follows from (2.18b) that the correlation function of the received process can be expressed as

$$R_y(t, \tau) = \iint \tilde{R}_{\alpha\beta}(t, \tau) \, d\alpha \, d\beta, \tag{2.19}$$

where $\tilde{R}_{\alpha\beta}(t, \tau)$ is the cross correlation function between $v_\alpha(t)$ and $v_\beta(t)$. In particular, for the point-scatterer model of Section 2.1, it is easy to show that

$$\tilde{R}_{\alpha\beta}(t, \tau) = E_r \text{ Re } [R_{\alpha\beta}(t, \tau) \exp j\omega_0(t - \tau)], \tag{2.20a}$$

where

$$R_{\alpha\beta}(t, \tau) = \iint \sigma(r, f)u_\alpha(t - r)u_\beta^*(\tau - r) \exp j\omega(\tau - t) \, dr \, df. \tag{2.20b}$$

We next determine the complex correlation functions of (2.20b) when $u(t)$ is expressed in terms of its frequency components. Specifically we represent $u(t)$ in the form

$$u(t) = \int U(f) \exp j2\pi ft \, df, \tag{2.21a}$$

where $U(f)$ is the Fourier transform of $u(t)$, that is,

$$U(f) = \int u(t) \exp -j2\pi ft \, dt. \tag{2.21b}$$

For this representation $u_\alpha(t)$ is $U(\alpha) \exp j2\pi\alpha t$, and (2.20b) becomes

$$R_{\alpha\beta}(t, \tau) = U(\alpha)U^*(\beta)\mathcal{R}(\beta - \alpha, \tau - t) \exp j2\pi(t\alpha - \tau\beta), \qquad (2.22a)$$

where $\mathcal{R}(\cdot, \cdot)$ is the double Fourier transform of $\sigma(r, f)$, that is,

$$\mathcal{R}(\Omega, t) = \iint dr \, df \, \sigma(r, f) \exp j2\pi(r\Omega + ft). \qquad (2.22b)$$

The function $\mathcal{R}(\cdot, \cdot)$ is usually called the *two-frequency correlation function* of the channel because of its role in determining the correlation between the response to two sinusoids.† Its properties are discussed elsewhere [15–19].

Upon introducing (2.22a) into (2.20a) and then into (2.19) and rearranging terms, we obtain

$$R_y(t, \tau) = E_r \, \text{Re} \, [R(t, \tau) \exp j\omega_o(t - \tau)] \qquad (2.23a)$$

with

$$R(t, \tau) = \iint U(\alpha)U^*(\beta)\mathcal{R}(\beta - \alpha, \tau - t) \exp j2\pi(\alpha t - \beta\tau) \, d\alpha \, d\beta. \quad (2.23b)$$

For the point-scatterer model of Sections 2.1 and 2.2, (2.23b) can be derived more simply by introducing the transform of $u(t)$ into (2.16). However, the derivation presented here does not really depend upon the existence of any point scatterers; rather, it depends on two assumptions pertaining to the transmission properties of the medium. First, the medium is assumed to be linear. Second, it is assumed that the cross correlation function between the responses to two sinusoids of frequencies α and β can be expressed as

$$\text{Re} \, \{U(\alpha)U^*(\beta)\mathcal{R}(\beta - \alpha, \tau - t) \exp j[\omega_o(t - \tau) + 2\pi(r\Omega + ft)]\},$$

where $\mathcal{R}(\cdot, \cdot)$ is of the form specified by (2.22b). That is, $\mathcal{R}(\cdot, \cdot)$ is the double Fourier transform of a nonnegative function and $\mathcal{R}(0, 0) = 1$.

The scatterer model of Section 2.1 certainly satisfies the aforestated assumptions, but the supposition that such scatterers exist is unnecessary. We may simply choose to study the class of channels that satisfy the assumptions. These channels, which are called wide-sense stationary uncorrelated scattering (WSSUS) channels, are known to provide useful models for many physical communication channels [19]. Of course, we may envisage any given WSSUS channel as though it is the scattering channel described by the scattering function

$$\sigma(r, f) = \iint \mathcal{R}(\alpha, \beta) \exp -j2\pi(\alpha r + \beta f) \, d\alpha \, d\beta, \qquad (2.24)$$

regardless of whether the real channel involves any scattering. The description

---

† The term time-frequency correlation function is also used [19]. Some definitions replace f by $-f$ in the exponent of the integrand.

of the channel is complete if the output is known to be a zero-mean Gaussian random process.

An alternate, and more compact, form of (2.23b) is obtained by introducing the two-dimensional correlation function of $u(t)$ [20]. This function is denoted by $\theta(\tau, f)$ and is defined by the expression

$$\theta(\tau, f) = \int u\left(t - \frac{\tau}{2}\right) u^*\left(t + \frac{\tau}{2}\right) \exp j2\pi f t \, dt \qquad (2.25a)$$

or equivalently

$$\theta(\tau, f) = \int U\left(\Omega - \frac{f}{2}\right) U^*\left(\Omega + \frac{f}{2}\right) \exp -j2\pi\Omega\tau \, d\Omega, \qquad (2.25b)$$

where $U(f)$ is again the Fourier transform of $u(t)$.†

To obtain the simpler form, we introduce new variables of integration $[f = \beta - \alpha$ and $x = (\beta + \alpha)/2]$ into (2.23b), eliminate α and β, and regroup terms. The result is

$$R_y(t, \tau) = E_r \, \text{Re} \, [R(t, \tau) \exp j\omega_o(t - \tau)] \qquad (2.26a)$$

with

$$R(t, \tau) = \int \mathcal{R}(f, \tau - t) \, \theta(\tau - t, f) \exp -j\pi f(t + \tau) \, df. \qquad (2.26b)$$

At this juncture we have developed our channel model from the concepts of linear-system theory as well as from the point-scatterer approach. We now briefly relate the model employed here to another, more general, one. Since this more general model is discussed thoroughly elsewhere, we limit our discussion to the correspondence between the two models [19, 22–24]. This correspondence is not used in the sequel, so Section 2.4 may be omitted without loss of continuity.

🦕 2.4 DELAY-LINE MODEL

In the delay-line model the received process is expressed in the form

$$y(t) = \text{Re} \left[\sum_i^k m_i(t) u(t - \Delta_i) \exp j\omega_o(t - \Delta_i) \right], \qquad (2.27)$$

where the $m_i(t)$ are complex functions. Thus the channel may be represented by a tapped delay line as shown in Fig. 2.3.

---

† The symbol $\chi(\tau, f)$ is sometimes used in place of $\theta(\tau, f)$ [19]. The magnitude squared $|\theta(\tau, f)|^2$ of $\theta(\tau, f)$ is the familiar radar ambiguity function [15, 20, 21].

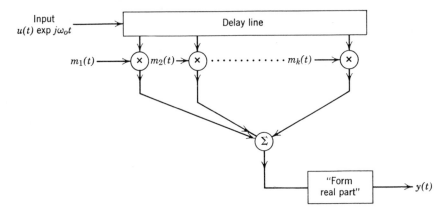

Fig. 2.3   Delay-line model for complex inputs.

As illustrated in that figure, $u(t) \exp j\omega_o t$ is fed into the delay line, and the waveform emerging from each tap is multiplied by a *complex tap-gain function* $m_i(t)$. The resulting waveforms are then added together, and their real part is extracted to obtain the channel-output waveform $y(t)$.

The mathematical model of Fig. 2.3 may be used to describe a wide variety of physical channels. However the utility of the model is limited by its mathematical tractability, for example, by the nature of the tap-gain functions. For fading dispersive channels one frequently assumes that these functions, and hence $y(t)$, are zero-mean Gaussian random processes. This implies that the statistical properties of $y(t)$ are completely specified by the correlation function

$$R_y(t, \tau) = \overline{y(t)y(\tau)}. \tag{2.28}$$

An expression for $R_y(t, \tau)$ may be obtained by using (2.28) but it is rather cumbersome [19]. Therefore in the interests of simplicity and also because of limited knowledge of the channel, one often further assumes that the gain functions associated with different taps are uncorrelated stationary random processes. The final common assumption is that the real and imaginary parts of each $m_i(t)$ are uncorrelated and possess the same correlation function. This implies that the expected value of $m_i(t)m_i(t + \tau)$ vanishes for all i and τ.

The preceding assumptions reduce the general model of Fig. 2.3 to that of Fig. 2.2. To demonstrate this, we first note that, by virtue of the assumptions,

$$R_y(t, \tau) = \mathrm{Re}\left[\frac{1}{2} \sum_{i=1}^{k} G_i(t - \tau)u(t - \Delta_i)u^*(\tau - \Delta_i) \exp j\omega_o(t - \tau) \right], \tag{2.29a}$$

where $G_i(\tau)$ is called the ith tap-gain correlation function, that is,

$$G_i(\tau) = \overline{m_i(t)m_i^*(t - \tau)}. \tag{2.29b}$$

An alternate and more convenient expression may be obtained either by defining

$$G(r, t - \tau) = \sum_{i=1}^{k} \delta(r - \Delta_i)G_i(t - \tau), \qquad (2.30)$$

where $\delta(r)$ is the unit impulse, or by assuming that the tap spacing is much less than the reciprocal bandwidth of the transmitted waveform so that the summation of (2.29a) may be regarded as an integral. In either event

$$R_y(t, \tau) = \frac{1}{2} \text{Re} \left[\int G(r, t - \tau)u(t - r)u^*(\tau - r) \exp j\omega_o(t - \tau) \, dr \right]. \quad (2.31)$$

We next observe that the model of Fig. 2.2 yields the correlation function of (2.31) if the scattering function and average energy are chosen to be

$$\sigma(r, f) = \frac{1}{2E_r} \int G(r, x) \exp j2\pi xf \, dx \qquad (2.32)$$

and

$$E_r = \frac{1}{2} \int G(r, 0) \, dr, \qquad (2.33)$$

respectively. These choices are admissable because the resulting E_r and $\sigma(r, f)$ are nonnegative.† Thus we conclude that the conditions imposed upon the tap-gain functions do indeed reduce the model of Fig. 2.3 to that of Fig. 2.2.

As a note of explanation, we have included a discussion of the delay-line model because of its generality and because it may provide a point of reference for many communication engineers. On the other hand, we choose to employ the point-scatterer description as our basic model because it is particularly useful for many channels of interest and because it involves suppositions that are more physically motivated than those pertaining to the tap-gain correlation functions are. However, many of our subsequent results apply to any channel for which the output is a zero-mean Gaussian process, even if it is not a WSSUS channel.

An alternate channel description, which in a sense is dual to Fig. 2.3, can be obtained from a delay-line model that operates upon the transform of the transmitted waveform rather than upon the waveform itself. The specification of this model involves a (frequency) tap-gain correlation function that is the Fourier transform of the scattering function with respect to the range

---

† That they are nonnegative is an immediate consequence of the assumption that each $m_i(t)$ is stationary with uncorrelated real and imaginary parts.

variable rather than with respect to the frequency variable. We do not pursue its development here since it is discussed extensively elsewhere [19, 22–24].

2.5   DIVERSITY REPRESENTATION

Thus far we have concentrated upon the correlation function of the received process. Although this function does describe the system, subject to our zero-mean Gaussian assumption, it provides little insight into the structure of the received process. Therefore we now introduce a representation that depicts a fading dispersive channel as a diversity system. This representation is used repeatedly in subsequent chapters.

The results of this section are, in essence, applications of the Fredholm theory of integral equations and of the Karhunen-Loève orthogonal expansion theorem. We employ, without proof, the relevant results of these theories; the proofs can be found in the literature [25–28].

Karhunen-Loève Expansion

The Karhunen-Loève theorem states that, subject to a condition described below, a random process may be expanded in an orthogonal series with uncorrelated coefficients. Specifically, if $y(t)$ is a sample function of a real process whose mean is zero and whose correlation function is $R_y(t, \tau)$, then $y(t)$ is expressible in the form

$$y(t) = \sum_{i=1} y_i \, \tilde{\varphi}_i(t), \qquad (2.34a)$$

where

$$y_i = \int y(t) \tilde{\varphi}_i^*(t) \, dt. \qquad (2.34b)$$

The sense of the "equality" in (2.34a) is discussed subsequently.

The functions $\tilde{\varphi}_i(t)$ are the normalized *eigenfunctions* of $R_y(t, \tau)$. That is, each $\tilde{\varphi}_i(t)$ satisfies the integral equation

$$\int R_y(t, \tau) \tilde{\varphi}_i(\tau) \, d\tau = \tilde{\lambda}_i \, \tilde{\varphi}_i(t), \qquad (2.35)$$

where $\tilde{\lambda}_i$ is the *eigenvalue* associated with $\tilde{\varphi}_i(t)$. Conversely, every solution of (2.35) is expressible as a linear combination of the $\tilde{\varphi}_i(t)$. One important property of these eigenfunctions is that they can be chosen to be orthonormal, that is,

$$\int \tilde{\varphi}_i(t) \tilde{\varphi}_j^*(t) \, dt = \begin{cases} 0, & \text{if } i \neq j, \\ 1, & \text{if } i = j. \end{cases} \qquad (2.36)$$

The coefficients y_i appearing in the Karhunen-Loève expansion of (2.34) are zero-mean uncorrelated random variables with mean square λ_i; that is,

$$\overline{y_i} = 0,$$
$$\overline{y_i y_j} = 0, \quad \text{for } i \neq j, \tag{2.37}$$
$$\overline{y_i^2} = \tilde{\lambda}_i.$$

Moreover when $y(t)$ is a Gaussian random process, the coefficients are Gaussianly distributed random variables. In that event they are not only uncorrelated but also statistically independent of each other.

Loosely stated, the condition associated with (2.34) is that the correlation function $R_y(t, \tau)$ of the process be square integrable over the time interval for which the expansion is made. We are concerned here only with the doubly infinite interval so the condition is that

$$\iint [R_y(t, \tau)]^2 \, dt \, d\tau < \infty. \tag{2.38}$$

Although (2.38) is not satisfied for all correlation functions, it is satisfied for all those that can occur in our model of fading dispersive channels. Indeed it is easy to show that, for correlation functions of the form given in Fig. 2.2, the integral of (2.38) does not exceed E_r^2.

To render the loose statement of the preceding paragraph precise, we must consider the sense in which the right member of (2.34a) represents, or converges to, the left member. We choose not to become enmeshed in that question here. Suffice it to say that the convergence is at least in the "integrated-mean-square" sense; that is,

$$\lim_{N \to \infty} \int \overline{\left[y(t) - \sum_{i=1}^{N} y_i \tilde{\phi}_i(t) \right]^2} \, dt = 0.$$

If the finite-sum assumption of (2.8) is invoked, the convergence is "almost everywhere" for every sample function of finite energy because there are then only a finite number of terms in (2.34a).

Complex Representation

The expansion of (2.34) provides a description of the channel-output waveform as a sum of statistically independent "pieces." However, for our purposes, it is more convenient to replace, or approximate, this representation by one involving the eigenfunctions of the complex correlation function $R(t, \tau)$ associated with the complex envelope of $y(t)$. The replacement involves no approximation if the complex envelope or equivalently the quadrature components of the received waveform can be recovered from $y(t)$. Mathematically the supposition is that

$$\int R(t, s)R^*(s, \tau) \exp j2\omega_o s \, ds = 0, \qquad \text{for all } t, \tau. \qquad (2.39)$$

We now first establish the desired representation under the assumption that (2.39) is satisfied and then consider the significance of that equation.

First, we recall from (2.16a) that $R_y(t, \tau)$ may be expressed in the form

$$R_y(t, \tau) = E_r \, \text{Re} \, [R(t, \tau) \exp j\omega_o(t - \tau)] \qquad (2.40a)$$

or equivalently

$$R_y(t, \tau) = \frac{E_r}{2} [R(t, \tau) \exp j\omega_o(t - \tau)] + \frac{E_r}{2} [R^*(t, \tau) \exp -j\omega_o(t - \tau)], \qquad (2.40b)$$

where the complex correlation function $R(t, \tau)$ is given by the expression

$$R(t, \tau) = \int \sigma(r, f) \, u(t - r) \, u^*(\tau - r) \exp j\omega(\tau - t) \, dr \, df. \qquad (2.40c)$$

We next observe that $R(t, \tau)$ is a Hermitian kernel, that is,

$$R(t, \tau) = R^*(\tau, t). \qquad (2.41)$$

Therefore it can be expressed in the form

$$R(t, \tau) = \sum_{i=1} \lambda_i \, \varphi_i(t) \, \varphi_i^*(\tau), \qquad (2.42)$$

where the λ_i and $\varphi_i(t)$ are, respectively, the eigenvalues and normalized eigenfunctions of $R(t, \tau)$. Precisely stated, the right member of (2.42) converges to the left member in the integrated-mean-square sense, and this can be strengthened to an almost-everywhere convergence by invoking the finite-sum assumption of (2.8).

Upon introducing the right member of (2.42) into the right member of (2.40b), we obtain

$$R_y(t, \tau) = E_r \sum_{i=1} \left[\frac{\lambda_i}{2} \psi_i(t) \, \psi_i^*(\tau) + \frac{\lambda_i}{2} \psi_i^*(t) \, \psi_i(\tau) \right], \qquad (2.43a)$$

where we define

$$\psi_i(t) = \varphi_i(t) \exp j\omega_o t \qquad (2.43b)$$

and exploit the fact that the λ_i are real.

The $\psi_i(t)$ and the $\psi_i^*(t)$ considered separately form orthonormal sets because the $\varphi_i(t)$ are orthonormal, that is,

$$\int \psi_i(t)\psi_j^*(t) \, dt = \int \varphi_i(t)\varphi_j^*(t) \, dt = \begin{cases} 0, & \text{if } i \neq j, \\ 1, & \text{if } i = j. \end{cases} \qquad (2.44)$$

Consequently the totality of the $\psi_i(t)$ and $\psi_i^*(t)$ form an orthonormal set if

$$\int \psi_i(t)\psi_j(t)\,dt = 0 \tag{2.45}$$

for all values of i and j. Moreover it is easy to verify that this integral vanishes for all values of i and j if, as we suppose, (2.39) is satisfied.

The orthogonality of the combined set of $\psi_i(t)$ and $\psi_i^*(t)$, in conjunction with (2.35) and (2.43a), implies that each of the $\psi_i(t)$, and also each of the $\psi_i^*(t)$, is an eigenfunction of $R_y(t, \tau)$. Precisely stated, if the eigenvalues of $R(t, \tau)$ are $\lambda_1, \lambda_2, \ldots$ and the corresponding eigenfunctions are $\varphi_1(t)$, $\varphi_2(t), \ldots$, then

(a) the eigenvalues of $R_y(t, \tau)$ are $E_r \lambda_i/2$, $i = 1, 2, \ldots$;
(b) the two functions $\varphi_i(t) \exp j\omega_0 t$ and $\varphi_i^*(t) \exp -j\omega_0 t$ are eigenfunctions associated with $E_r \lambda_i/2$, $i = 1, \ldots$.

These properties make it possible to restate the Karhunen-Loève expansion of (2.34) in the following form:

$$y(t) = \sqrt{\frac{E_r}{2}}\,[z(t) \exp j\omega_0 t + z^*(t) \exp -j\omega_0 t] \tag{2.46a}$$

or equivalently

$$y(t) = \sqrt{2E_r}\,\mathrm{Re}\,[z(t) \exp j\omega_0 t], \tag{2.46b}$$

where $z(t)$, the (normalized) complex envelope of $y(t)$, is

$$z(t) = \sum_i z_i \varphi_i(t) \tag{2.46c}$$

and where the (complex) coefficients z_i, and their conjugates z_i^*, are defined by the expression

$$z_i = \sqrt{\frac{2}{E_r}} \int y(t)\varphi_i^*(t) \exp -j\omega_0 t\,dt. \tag{2.46d}$$

Equations 2.46 provide the representation we seek; it also establishes the significance of (2.39), for (2.46), which are a consequence of (2.39), clearly indicate that the complex envelope $z(t)$ can be constructed from $y(t)$. Or, stated alternately, (2.39) implies that the quadrature components of $y(t)$ can be recovered from $y(t)$.

As for (2.39) and (2.45), they are approximately satisfied if the carrier frequency ω_0 is much greater than the bandwidth of the received waveforms. In fact, the condition of (2.45) can be expressed as

$$\int \varphi_i(t)\varphi_j(t) \exp j2\omega_0 t\,dt = 0, \tag{2.47}$$

and this condition is satisfied in the limit of increasing ω_0 for any reasonably well-behaved functions $\varphi_i(t)$ and $\varphi_j(t)$ [29]. Consequently we may suppose that (2.39) and (2.45) are satisfied to a rather good approximation if ω_0 is sufficiently large. Alternately (2.45) is precisely satisfied if the sample functions of the received process are bandlimited to within the frequency band $-2\omega_0$ to $+2\omega_0$, for then the $\varphi_i(t)$ are bandlimited to within $-\omega_0$ to $+\omega_0$, $\varphi_i(t)\varphi_j(t)$ is bandlimited to within $-2\omega_0$ to $+2\omega_0$, and the integral of (2.47), which is the Fourier transform of $\varphi_i(t)\varphi_j(t)$ evaluated at $2\omega_0$, vanishes.

The complex coefficients z_i of (2.46) possess several properties that are immediate consequences of the original Karhunen-Loève expansion of (2.34) and of the (approximate) properties of the $\varphi_i(t)$. We state them here for future reference.

First, the real, and also the imaginary, parts of the coefficients z_i are always zero-mean uncorrelated random variables. The real parts are uncorrelated with the imaginary parts, and the real and imaginary parts possess the same variance $\lambda_i/2$. Stated alternately,

$$
\begin{aligned}
\overline{z_i} &= 0, \\
\overline{z_i z_j} &= 0, \qquad \text{for all } i \text{ and } j, \\
\overline{z_i z_j^*} &= 0, \qquad \text{for } i \neq j, \\
\overline{z_i z_i^*} &= \overline{|z_i|^2} = \lambda_i,
\end{aligned}
\tag{2.48}
$$

where λ_i is, as before, the ith eigenvalue of the complex correlation function.

Second, the λ_i sum to unity; that is,

$$
\sum_i \lambda_i = 1. \tag{2.49}
$$

This statement follows from (2.40) and (2.42). Specifically

$$
\sum_i \lambda_i = \sum_i \left[\lambda_i \int \varphi_i(t)\, \varphi_i^*(t)\, dt \right] = \int R(t, t)\, dt = 1. \tag{2.50}
$$

Third, if the received waveform is a sample function of a Gaussian random process, the z_i are statistically independent Gaussian variates. This implies that the magnitude of each z_i is a Rayleigh-distributed random variable whose mean-square value is λ_i and that the phase of each z_i is independent of the magnitude and is uniformly distributed in the interval $-\pi$ to π; that is,

$$
z_i = \tilde{x}_i \exp j\tilde{\theta}_i, \tag{2.51a}
$$

where the joint probability density $p(\tilde{x}_i, \tilde{\theta}_i)$ of $\tilde{x}_i$ and $\tilde{\theta}_i$ is

$$
p(\tilde{x}_i, \tilde{\theta}_i) = \frac{\tilde{x}_i}{\pi \lambda_i} \exp - \frac{(\tilde{x}_i)^2}{\lambda_i}, \qquad \text{for } |\tilde{\theta}| < \pi \text{ and } \tilde{x}_i \geq 0,
$$

$$
p(\tilde{x}_i, \tilde{\theta}_i) = 0, \qquad\qquad \text{elsewhere.}
\tag{2.51b}
$$

The representation we seek is provided by (2.46) together with the properties summarized in (2.48) to (2.51). The received process is characterized by $z(t)$, which is a sum of known orthonormal time functions $\varphi_i(t)$, each weighted by one of the independent random variables z_i. This representation permits us to visualize any fading dispersive channel as though it is a classical diversity system, that is, as a system wherein the transmitted waveform consists of several constituent parts, each of which may be detected separately [30–38].

This interpretation is illustrated in Fig. 2.4, where the system is depicted as one in which all of the orthogonal signals

$$\sqrt{2E_r}\,\text{Re}\,[\varphi_i(t)\exp \omega_o t], \qquad i = 1, 2, \ldots,$$

are transmitted simultaneously. Each of these waveforms is then altered, as in many diversity systems, by a uniformly distributed carrier phase shift $\tilde{\theta}_i$ and by a Rayleigh-distributed gain (or fade) $\tilde{x}_i$ of mean-square value λ_i. Thus we sometimes call λ_i the fraction of the total average received energy contributed by the ith *diversity path*, or, for brevity, the *fractional path strength*.

The representation of Fig. 2.4 is easily generalized to channels for which the output process is not Gaussian. In fact, the Gaussian assumption is invoked only to state that the z_i are statistically independent of each other and that the magnitude and phase of the z_i are, respectively, Rayleigh- and uniformly distributed random variables. Although these statements are of crucial importance in determining the form of the optimum detector, they are often of lesser importance in determining the performance of that detector. Moreover we can show that many channels involving large numbers

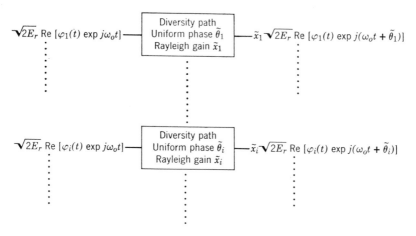

Fig. 2.4   Diversity representation of fading dispersive channel.

of point scatterers with negligible multiple scattering satisfy the Gaussian supposition. In particular, it is shown in Appendix 1 that the z_i become statistically independent Rayleigh-amplitude uniform-phase random variables with variances λ_i if the number of scatterers tends to infinity in such a way that the (fractional) cross section per scatterer vanishes.

It is important to note that our representation of the received process is conditioned upon the knowledge of the transmitted waveform. When we consider communication systems (Chapter 4), a representation that is valid for all the waveforms in a set is required. However the complexities of such representations for general sets of waveforms ultimately leads us back to the representation considered here.

It should also be noted that, insofar as systems analysis is concerned, there is no essential difference between a true diversity system and a fading dispersive channel. Both are described by the set of transmitted waveforms (or eigenfunctions) and the set of corresponding average path energies (or eigenvalues). However a crucial difference arises in the design of communication systems. The designer of a diversity system may specify both the number of diversity paths and the transmitted waveforms, whereas the designer of a fading dispersive system can specify only the complex envelope $u(t)$ that serves to determine the eigenfunctions and eigenvalues of the complex correlation function. Thus one of our eventual tasks is to ascertain the dependence of the $\varphi_i(t)$ and of the λ_i upon the waveform employed.

This task is a formidable one because of the difficulty in determining the eigenvalues of a given complex correlation function. Therefore we supplement our precise results with oversimplified, but useful, descriptions employing the gross properties of the channel scattering function and of the complex envelope. These descriptions are discussed in Chapter 3. In that, and subsequent, discussions we occasionally employ an equivalence described in Section 2.6.

2.6 EQUIVALENT CHANNELS

Although it is usually difficult to determine the eigenvalues and eigenfunctions of a complex correlation function, there are some transformations that relate the solution of a single problem to a whole class of problems. These transformations permit us to conclude that the channels whose scattering functions are of a particular form are, in a sense, equivalent.

To illustrate the character of the equivalence, let us consider two channels with scattering functions

$$\sigma_1(r, f) = g(r, f) \qquad (2.52a)$$

and

$$\sigma_2(r, f) = g(f, -r) \tag{2.52b}$$

and with complex inputs $u_1(t)$ and $u_2(t)$. We claim that the outputs of these two channels are closely related when $u_2(t)$ is the Fourier transform of $u_1(t)$. Specifically the complex envelope of the output of the second channel is the Fourier transform of the complex envelope of the first channel's output. Moreover the complex correlation functions of the two systems possess the same set of eigenvalues, and the eigenfunctions are Fourier transforms of each other.

To verify the claim we first show that the complex correlation functions of the two systems are Fourier transforms of each other. To this end, we employ (2.26b) to obtain

$$R_1(t, \tau) = \int \mathscr{R}_1(\alpha, \tau - t)\theta_1(\tau - t, \alpha) \exp -j\pi\alpha(t + \tau) \, d\alpha \tag{2.53a}$$

and

$$R_2(t, \tau) = \int \mathscr{R}_2(\alpha, \tau - t)\theta_2(\tau - t, \alpha) \exp -j\pi\alpha(t + \tau) \, d\alpha, \tag{2.53b}$$

where $\mathscr{R}_1(\cdot, \cdot)$ and $\mathscr{R}_2(\cdot, \cdot)$ are the two-frequency correlation functions associated with $\sigma_1(\cdot, \cdot)$ and $\sigma_2(\cdot, \cdot)$ and where $\theta_1(\cdot, \cdot)$ and $\theta_2(\cdot, \cdot)$ are the two-dimensional correlation functions associated with $u_1(t)$ and $u_2(t)$ [see (2.25)].

We note from (2.22) and (2.25) that, for the special problem we are considering, the two-dimensional correlation functions and the two-frequency correlation functions are related by the expressions

$$\theta_2(\beta, \alpha) = \theta_1(-\alpha, \beta) \tag{2.54a}$$

and

$$\mathscr{R}_2(\alpha, \beta) = \mathscr{R}_1(\beta, -\alpha). \tag{2.54b}$$

Upon introducing the right members of these equations into (2.53b) and taking the mixed double Fourier transform of the result, we obtain, after some manipulation,

$$\iint R_2(t, \tau) \exp j2\pi(tx - \tau y) \, dt \, d\tau = R_1(x, y) \tag{2.55}$$

as was to be shown.

We next claim that $R_1(t, \tau)$ and $R_2(t, \tau)$ possess the same set of eigenvalues and possess eigenfunctions that are Fourier transforms of each other. This is so, for if

$$\int R_1(t, \tau) \, \varphi_i(\tau) \, d\tau = \lambda_i \varphi_i(t) \tag{2.56a}$$

then by straightforward manipulation of integrals

$$\int R_2(t, \tau)\, \hat{\varphi}_i(\tau)\, d\tau = \lambda_i\, \hat{\varphi}_i(t), \qquad (2.56b)$$

where

$$\hat{\varphi}_i(t) = \int \varphi_i(\tau) \exp -j2\pi t\tau\, d\tau. \qquad (2.56c)$$

It is also easy to show that, if (2.56b) is satisfied, so also is (2.56a) with $\varphi_i(t)$ being the inverse Fourier transform of $\hat{\varphi}_i(t)$. Consequently $R_1(t, \tau)$ and $R_2(t, \tau)$ possess the same set of eigenvalues, and the eigenfunctions of $R_2(t, \tau)$ are the Fourier transforms of the eigenfunctions of $R_1(t, \tau)$, as claimed.

To show that the complex envelopes of the outputs of the two channels are related through the Fourier transformation, we employ the expansion of (2.46c). According to that equation and to the results of the preceding paragraph, the expansions for the two complex envelopes differ only in that the orthonormal functions involved in one expansion are the Fourier transforms of those involved in the other. Since the expansion is linear, this transformation may be applied to the first expansion to obtain the second; that is, the complex random process $\hat{z}(t)$ associated with the output of the second channel is identical to the random process obtained by Fourier transforming the complex process $z(t)$ associated with the first channel's output.

We may summarize the results of this simple example as follows: the eigenvalues of a system are invariant to a simultaneous 90° counterclockwise rotation of the scattering function and 90° clockwise rotation of the two-dimensional correlation function. The complex envelope of the channel-output process is carried into its Fourier transform by this rotation.

The property described above is an example of the time-frequency duality discussed elsewhere [19, 39]; we have occasion to use it in Chapter 3. It is also an example of a broader class of transformations that do not alter the eigenvalues associated with a transmitted waveform and a channel scattering function. Such invariance, or equivalence, classes are important to us because they broaden the class of channels to which our subsequent specific results apply. The proof of this equivalence is omitted since it is similar to that employed in the example; the conclusions are as follows:

We define two channels to be equivalent if, for every possible transmitted waveform used with one of them, there exists a waveform that can be used with the other so as to obtain identical sets of eigenvalues in the two systems; for example, the channels described by (2.52a) and (2.52b) are equivalent because the eigenvalues associated with the two systems are identical when one transmitted waveform is the Fourier transform of the other.

More generally, all channels whose scattering functions are of the form

$$\sigma(r, f) = g\left(ar - \frac{cf}{a}, akr + \frac{1 - kc}{a}f\right) \qquad (2.57)$$

are equivalent no matter what the values of c, k, and a. In particular, if the channel with the scattering function

$$\tilde{\sigma}(r, f) = g(r, f) \qquad (2.58a)$$

is employed with a waveform whose complex envelope is

$$\tilde{u}(t) = \sqrt{a} \iint u(\alpha) \exp j\pi(2a\alpha\beta - c\beta^2 - 2\beta t - kt^2) \, d\beta \, d\alpha, \qquad (2.58b)$$

the eigenvalues of the system will be identical to those obtained with a scattering function

$$\sigma(r, f) = g\left(ar - \frac{c}{a}f, akr + \frac{1 - kc}{a}f\right) \qquad (2.59)$$

and a waveform whose complex envelope is $u(t)$. Moreover the complex-

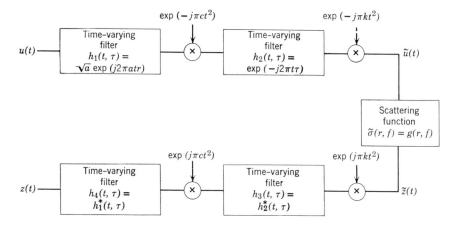

Fig. 2.5 Equivalent channels.

output random process $z(t)$ of the latter system is related to that, $\tilde{z}(t)$, of the former system through the expression

$$z(t) = \sqrt{a} \iint \tilde{z}(\alpha) \exp j\pi(k\alpha^2 + 2\alpha\beta + c\beta^2 - 2a\beta t) \, d\beta \, d\alpha. \qquad (2.60)$$

These relations are summarized in Fig. 2.5. As a specific example of this result, we set

$$a = \cos \varphi,$$

$$k = \tan \varphi,$$

$$c = \sin \varphi \cos \varphi,$$

where φ is an arbitrary angle. Substitution of these values into (2.57) yields

$$g(r \cos \varphi - f \sin \varphi, r \sin \varphi + f \cos \varphi). \qquad (2.61)$$

Since this expression is just a clockwise rotation of $g(r, f)$ through an angle φ, we conclude that all channels whose scattering functions are rotations of a given scattering function are equivalent to each other. An example of equivalence under dilation is obtained by setting k and c equal to zero which leads us to conclude that all scattering functions of the form $\sigma(ar, f/a)$ are equivalent.

REFERENCES

[1] D. J. Sakrison, *Communication Theory: Transmission of Waveforms and Digital Information.* New York: Wiley, 1968, p.132.

[2] C. W. Helstrom, *Statistical Theory of Signal Detection.* New York: Pergamon, 1960, p. 11.

[3] M. Schwartz, W. R. Bennett, and S. Stein, *Communication Systems and Techniques.* New York: McGraw-Hill, 1966, p. 29.

[4] J. Dugundji, "Envelopes and Pre-Envelopes of Real Waveforms." *IRE Trans. Inform. Theory*, 53–57, March 1958.

[5] R. Arens, "Complex Processes for Envelopes of Normal Noise." *IRE Trans. Inform. Theory*, 204–207, September 1957.

[6] M. Schwartz, W. R. Bennett, and S. Stein, *Communication Systems and Techniques.* New York: McGraw-Hill, 1966, p. 364

[7] U.S. Department of Commerce, *Ionospheric Radio Propagation*, 1965, pp. 242 and 348.

[8] Y. L. Alpert, *Radio Wave Propagation and The Ionosphere.* New York: Consultants Bureau, 1963.

[9] R. A. Silverman and M. Balser, "Statistics of Electromagnetic Radiation Scattered by a Turbulent Medium." *Phys. Rev.* 560–563, November 1, 1954.

[10] M. Schwartz, W. R. Bennett, and S. Stein, *Communication Systems and Techniques*, New York: McGraw-Hill, 1966, pp. 347–397.

[11] J.T.A.C., "Radio Transmission by Ionospheric and Tropospheric Scatter." *Proc. IRE*, 4–29, January 1960.

[12] "Scatter Issue", *Proc. IRE*, October 1955.

[13] R. W. E. McNicol, "The Fading of Radio Waves of Medium and High Frequencies." Radio Section, No. 891, 517–524, October 1949.

[14] H. N. Shaver, B. C. Tupper, and J. B. Lomax, "Evaluation of a Gaussian HF Channel Model." *IEEE Trans. Commun. Technology*, 79–88, February 1967.

[15] J. W. Evans and T. Hagfors, Eds., *Radar Astronomy.* P. E. Green. New York: McGraw-Hill, 1968, Chapter 1, pp. 1–77.

[16] T. Hagfors, "Some Properties of Radio Waves Reflected from the Moon and their Relation to the Lunar Surface." *J. Geophysics Res.* **66**, 777–785, 1961.

[17] T. Kailath, "Measurements on Time-Variant Communication Channels." *IEEE Trans. Inform. Theory*, 229–236, September 1962.

[18] T. Kailath, "Time-Variant Communication Channels." *IEEE Trans. Inform. Theory*, 233–237, October 1963.

[19] P. A. Bellow, "Characterization of Randomly Time-Variant Linear Channels." *IEEE Trans. Commun. Systems*, 360–393, December 1963.

[20] W. M. Siebert, "Studies of Woodward's Uncertainty Function." Quarterly Progress Report, Research Laboratory of Electronics, M.I.T., 90–94, April 15, 1958.

[21] P. M. Woodward, *Probability and Information Theory, with Applications to Radar.* New York: Pergamon, 1953, pp. 118–125.

[22] J. C. Hancock and P. A. Wintz, *Signal Detection Theory.* New York: McGraw-Hill, 1966, pp. 16–30.

[23] E. J. Baghdady, Ed., *Lectures on Communication System Theory.* T. Kailath. New York: McGraw-Hill, 1961. Chapter 6, pp. 95–123.

[24] L. A. Zadeh, "Time-Varying Networks I." *Proc. IRE*, 1488–1502, October 1961.

[25] R. Courant and D. Hilbert, *Methods of Mathematical Physics.* New York: Interscience, Vol. I, 1953, Chapter 3.

[26] F. G. Tricomi, *Integral Equations.* New York: Interscience, 1957.

[27] F. Smithies, *Integral Equations.* London: Cambridge University Press, No. 49, 1958.

[28] W. B. Davenport, Jr., and W. L. Root, *Random Signals and Noise.* New York: McGraw-Hill, 1958, pp. 96–101.

[29] E. C. Titchmarsh, *The Theory of Functions.* London: Oxford University Press, 1939, p. 403.

[30] J. M. Wozencraft and I. M. Jacobs, *Principles of Communication Engineering.* New York: Wiley, 1965, pp. 529–535.

[31] J. C. Hancock and P. A. Wintz, *Signal Detection Theory.* New York: McGraw-Hill, 1966, pp. 191–198.

[32] E. J. Baghdady, Ed., *Lectures on Communication System Theory.* E. J. Baghdady, "Diversity Techniques." New York, McGraw-Hill, 1961, p. 125.

[33] G. F. Montgomery, "Message Error in Diversity Frequency-Shift Reception." *Proc. IRE*, 1184–1187, July 1954.

[34] J. N. Pierce, "Theoretical Diversity Improvement in Frequency-Shift Keying." *Proc. IRE*, 903–910, May 1958.

[35] D. G. Brennan, "Linear Diversity Combining Techniques." *Proc. IRE*, 1075–1102, June 1959.

[36] G. L. Turin, "On Optimal Diversity Reception." *IRE Trans. Inform. Theory*, 154–166, July 1961.

[37] G. L. Turin, "On Optimal Diversity Reception, II." *IRE Trans. Commun. Systems*, 22–31, March 1962.

[38] W. C. Lindsey, "Error Probability for Incoherent Diversity Reception." *IEEE Trans. Inform. Theory*, 491–499, October 1965.

[39] P. A. Bello, "Time-Frequency Duality." *IEEE Trans. Inform. Theory*, 18–33, January 1964.

[40] R. G. Gallager "Characterization and Measurement of Time- and Frequency-Spread Channels." M.I.T. Lincoln Laboratory Technical Report 352, Defense Documentation Center AD 443715, April 30, 1964.

[41] P. A. Bello, "On the Measurement of a Channel Correlation Function." *IEEE Trans. Inform. Theory*, 229–239, September 1962.

[42] J. J. Spilker, "On the Characterization and Measurement of Randomly Varying Filters." *IEEE Trans. Circuit Theory*, 44–51, March 1965.

[43] W. L. Root, "On the Measurement and Use of Time-Varying Communication Channels." *Information and Control*, 390–422, August 1965.

[44] P. A. Bello, "Some Techniques for the Instantaneous Real-Time Measurement of Multipath and Doppler Spread." *IEEE Trans. Commun. Technology*, 285–292, September 1965.

[45] I. Bar-David, "Estimation of Linear Weighting Functions in Gaussian Noise." *IEEE Trans. Inform. Theory*, 395–407, May 1968.

Chapter 3

Channel Characteristics

Our intent in this chapter is to lend additional substance to the mathematical model of fading dispersive channels. To this end, we first describe the scattering functions of some representative channels and discuss the classification of channels according to the character of the distortion associated with them. We then examine the distribution of the average received power in the time and the frequency domains and also establish some estimates of the coherence time and the coherence bandwidth of the received waveform. Finally we introduce a rough measure of the number of constituent diversity paths composing a fading dispersive system. These measures form the connecting link between the heuristic and approximate descriptions of this chapter and the precise but complex formulation of the preceding chapter.

For our purposes a scattering function may be translated from one region of the range-delay Doppler-shift plane to any other region without altering the relevant properties of the channel. This is so because such a translation merely changes the nominal time delay and carrier offset between the transmitted and received waveforms. Stated alternately, it corresponds to a change in the origin of the time and frequency scales at the receiver. Such changes are easily accounted for and need not concern us here. Thus we usually assume that the scattering function is translated so that the nominal time delay and carrier offset are zero.

3.1 REPRESENTATIVE CHANNELS

The scattering functions of several channels are illustrated in Figs. 3.1 through 3.5. That of Fig. 3.1 is the estimated scattering function for the F layer of the ionosphere at HF [1]. It shows the presence of three distinguishable "paths," or clusters of scatterers, each of which is composed of scatterers that are spread over a range delay interval of approximately $100 \mu s$ and a Doppler interval of approximately 0.1 to 1.0 Hz.

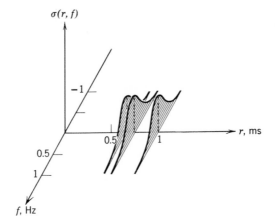

Fig. 3.1   Approximate scattering function for F-layer multipath.

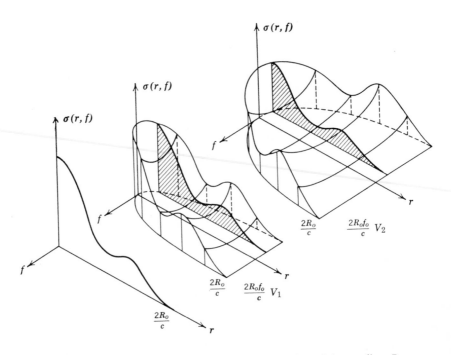

Fig. 3.2   Scattering function for uniform rough rotating sphere. Sphere radius: R_o; propagation velocity: c; carrier frequency: f_o; projected rotation velocity: V_1, V_2.

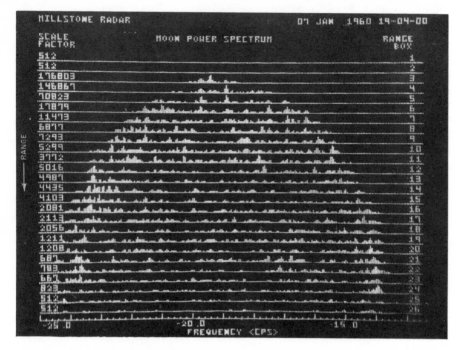

Fig. 3.3 Scattering function of the moon at 400 MHz.

The second scattering function (Fig. 3.2) is that resulting from an analysis of a uniformly rough rotating sphere [2]. The three parts of the figure pertain to different rotational velocities of the sphere. If this analysis is applied to the moon at a carrier frequency of 400 MHz, we find that the maximum widths of the scattering function in time and frequency are approximately 10 ms and 20 Hz.

For purposes of comparison, Fig. 3.3 presents the values of $\sigma(r, f)$ that resulted from a moon-mapping experiment [3]. In the figure, each range box represents approximately 400 μs of range delay and the range increases as we read down the range-box scale. The scale factors appearing in the figure are simply normalization constants employed to reduce the data to a presentable form. Thus, for example, the value of $\sigma(r, f)$ in range box 3 is on the order of 176,803/5,299 times larger than the value in range box 10. The gross similarities between Figs. 3.2 and 3.3 are obvious.

Some sections of another scattering function of interest are illustrated in Fig. 3.4. It is the scattering function of the orbiting dipole belt (Project West Ford) that was created in May, 1963 [4]. The figure, which is based upon measurements taken on May 20, 1963, clearly demonstrates the presence of

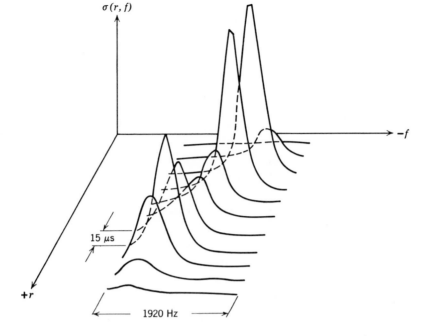

Fig. 3.4 Scattering function of West Ford dipole belt, May 1963.

two clusters of dipoles separated in range delay by approximately 60 μs and in Doppler by approximately 500 Hz. The width of each of the clusters is roughly 30 μs and 250 Hz. The results of subsequent measurements taken after the belt was fully deployed are shown in Fig. 3.5. It is readily apparent that the dipoles were, at this juncture (June 19, 1963), more evenly distributed.

We have grossly described the scattering functions of Figs. 3.1 through 3.5 by the widths of their approximately disjoint pieces. Of course, such a description may omit many important attributes of the scattering function; for example, the separation between the "pieces" is often of fundamental importance. However since our objective is to obtain simple descriptions that complement the precise formulation of Chapter 2, we are quite willing to tolerate the limitations of these descriptions. In fact, we often assume that the scattering function consists of a single "piece" so that it may be characterized even more simply.

The role of the widths we have discussed is analogous to the usual role of waveform time duration and bandwidth: they often permit a simple but approximate statement of results which are otherwise too cumbersome to present. Their precise definition is not very important since they are used only

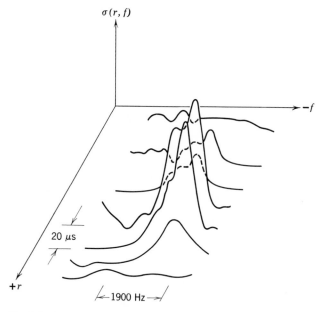

Fig. 3.5   Scattering function of West Ford dipole belt, June, 1963.

in an imprecise way. The definitions given in Section 3.2 are appropriate for our purposes.

3.2   MEASURE OF SPREAD

We grossly characterize scattering functions by their *Doppler spread B, multipath spread L,* and *total spread S.* Bearing in mind the limitations of the concepts, we adopt the following definitions [5]:

$$B = \left[\int \sigma^2(f)\, df \right]^{-1}, \tag{3.1a}$$

$$L = \left[\int \sigma^2(r)\, dr \right]^{-1}, \tag{3.1b}$$

and

$$S = \left[\iint \sigma^2(r, f)\, dr\, df \right]^{-1}, \tag{3.1c}$$

where

$$\sigma(r) = \int \sigma(r, f)\, df \tag{3.1d}$$

and†

$$\sigma(f) = \int \sigma(r, f) \, dr. \tag{3.1e}$$

A few remarks provide some interpretation of these definitions. First we recall that the value of $\sigma(r, f)$ may be regarded as the fraction of the scatterer cross section contributed by scatterers in the vicinity of range delay r and Doppler shift f. Thus $\sigma(r)$ is the fraction of cross section contributed by scatterers in the vicinity of r regardless of their Doppler shifts. Similarly $\sigma(f)$ is the fraction of the scatterer cross section contributed by scatterers whose Doppler shift is in the vicinity of f, regardless of their range delays. Alternately we may say that, if the Doppler spread of the scatterers "appears" to be zero in a given application, the scattering function reduces to $\sigma(r)$‡. Similarly if the multipath spread "appears" to be zero, the scattering function reduces to $\sigma(f)$. Thus it is natural to call $\sigma(r)$ the *delay scattering function* and $\sigma(f)$ the *frequency scattering function*.

The quantity B provides a rough measure of the frequency interval over which the Doppler shifts of the scatterers are spread, whereas L provides a measure of the interval over which their range delays are spread. For reasons that become apparent subsequently, these quantities are sometimes called the *frequency dispersion* and *time dispersion* of the channel. The remaining quantity S is a rough estimate of the area of the range-delay Doppler-shift plane over which the scatterers are spread. The values of B, L, and S for several scattering functions are given in Table 3.1.

It is evident from the table that these definitions do not reflect some crucial differences that can exist between different scattering functions. In fact, the three-parameter description tends to become useless if the scattering function consists of several disjoint pieces. Moreover the description can be quite ambiguous even when the scattering function is composed of a single piece. However the frequent simplification of results and concepts afforded by the introduction of these parameters more than offsets the caution that must be exercised in their use.

At this point it is convenient to define the duration and bandwidth of the transmitted waveform. As noted earlier, the exact definition is relatively unimportant. The definitions below suffice for our purposes.

Let the transmitted waveform $s(t)$ be represented in the complex form

$$s(t) = \text{Re} \, [u(t) \exp j\omega_o t]. \tag{3.2a}$$

---

† To be precise, we should distinguish the function $\sigma(r)$ from the function $\sigma(f)$ by more than a change of argument. However the meaning is always clear from the context and the additional precision is not worth the complication of the notation.

‡ Strictly speaking the scattering function reduces to $\delta(f - f_d)\sigma(r)$, where f_d is the common apparent Doppler shift of all the scatterers, but as noted in the introduction, f_d may be taken to be zero.

Table 3.1   Values of B, L, and S for Some Scattering Functions

$\sigma(r, f)$		Notes
$(BL)^{-1}$ for	$\begin{aligned} &|f| < \dfrac{B}{2} \\[4pt] &|r| < \dfrac{L}{2} \end{aligned}$	$S = BL$
$\dfrac{2}{BL} \exp - \dfrac{4|f|}{B}$ for $|r| < \dfrac{L}{2}$		$S = BL$
$\dfrac{4}{BL} \exp - 4\left(\dfrac{|f|}{B} + \dfrac{|r|}{L}\right)$		$S = BL$
$\dfrac{2}{S} \exp - \dfrac{2\pi}{S^2}\left[(rB)^2 + (fL)^2 - 2rf\sqrt{(BL)^2 - S^2}\right]$		$S \leq BL$
$\dfrac{2}{3BL}\left[1 + \cos\dfrac{4\pi nf}{3B}\right]$ for	$\begin{aligned} &|r| < \dfrac{L}{2} \\[4pt] &|f| < \dfrac{3B}{4} \end{aligned}$	$S = BL$ n any integer
S^{-1} for $r^2 + f^2 \leq \dfrac{S}{\pi}$		$B = L = \dfrac{3\pi}{16}\sqrt{\pi S}$

The *duration T* and *bandwidth W* of both $s(t)$ and also of $u(t)$ are defined by the expressions [5]

$$T = \left[\int |u(t)|^4 \, dt\right]^{-1} \tag{3.2b}$$

and

$$W = \left[\int |U(f)|^4 \, df\right]^{-1}, \tag{3.2c}$$

where $U(f)$ is the Fourier transform of $u(t)$, that is,

$$U(f) = \int u(t) \exp - j2\pi ft \, dt. \tag{3.2d}$$

It is evident from Table 3.2 that these definitions yield reasonable results when applied to waveforms that are concentrated in time and frequency. For segmented waveforms they tend to reflect the time (or frequency) over which the waveform is nonzero rather than the time (or frequency) from the "beginning" to the "end" of the waveform. The apparent difference between these definitions and those used for B, L, and S occurs because it is $|u(t)|^2$ that is analogous to $\sigma(r, f)$.

Table 3.2 Values of T and W for Some Waveforms

$u(t)$	TW product
$T^{-1/2} \quad$ for $\quad |t| < \dfrac{T}{2}$	1.5
$\sqrt{\dfrac{2}{T}} \exp - \dfrac{2|t|}{T}$	0.8
$\dfrac{2^{1/4}}{\sqrt{T}} \exp -\pi \left(\dfrac{t}{T}\right)^2 \left[1 + j\sqrt{(TW)^2 - 1}\right]$	≥ 1
$\sqrt{W} \dfrac{\sin \pi W t}{\pi W t}$	1.5

3.3 CLASSIFICATION OF CHANNELS

In the examples of Section 3.1 and in almost all realistic situations, neither the frequency dispersion B nor the time dispersion L vanishes. However the effects of one or both of these dispersions are sometimes indiscernable in the received waveform; that is, the system behaves as though either B or L or both are zero. Therefore it is instructive to consider the properties of channels for which either B or L or both are identically zero and to determine the conditions for which a system behaves as though one or both of them are zero.

Nondispersive Channels

We define a nondispersive, but fading, channel to be one whose scattering function is a unit impulse; that is,

$$\sigma(r, f) = \delta(r)\delta(f). \tag{3.3}$$

In accord with our earlier comments, we choose to locate the scattering function at the origin of the range-delay Doppler-shift plane. This channel is, in essence, the random-phase Rayleigh fading channel discussed extensively in the literature [6–10]. That is, the channel output $y(t)$ in response to the transmitted waveform

$$\text{Re}\left[u(t) \exp j\omega_o t\right] \tag{3.4a}$$

is

$$y(t) = A\rho \, \text{Re}\left[u(t) \exp j(\omega_o t - \theta)\right], \tag{3.4b}$$

where ρ and θ are statistically independent random variables that are not functions of time, the former being Rayleigh distributed and the latter being uniformly distributed in the interval $-\pi$ to π.

A nondispersive fading channel is sometimes called a *flat-flat fading* channel for reasons that become apparent subsequently [11]. As we see below, a channel appears to be nondispersive for a given transmitted waveform if both BT and LW are sufficiently small.

Channels Dispersive Only in Time

We say that a channel is dispersive only in time, or is not dispersive in frequency, if its scattering function is impulsive in frequency, that is, if

$$\sigma(r,f) = \delta(f)\sigma(r). \tag{3.5}$$

In this expression $\sigma(r)$ is the range-delay scattering function defined by (3.1d). Here we take the common Doppler shift of the scatterers to be zero.

Aside from the unimportant common Dopper shift, channels that are dispersive only in time compose linear time-invariant systems. This fact, which follows readily from (2.8), implies that the response of the channel to a sinusoid is the input sinusoid multiplied by a constant, albeit random, gain and delayed by a constant random phase; that is, the output does not appear to fade in time. For this reason channels that are dispersive only in time are occasionally called *time-flat fading* channels [11].

To gain some further understanding of the channel output process, we now suppose that the scatterers are confined to the delay interval $-L/2$ to $+L/2$, that is,

$$\sigma(r) = 0 \qquad \text{for} \quad |r| > \frac{L}{2}. \tag{3.6}$$

We also suppose that $u(t)$ is essentially zero for $|t| > T/2$ and that its transform $U(f)$ is essentially zero for $|f| > W/2$.

We recall that the received process is the superposition of the contributions from all the scatterers composing the medium. Therefore it includes components that have been subjected to (relative) range delays of $-L/2$ to $+L/2$, and the received process is spread over a time duration of approximately $T + L$ sec. Thus, if L is much less than T, there is no apparent spreading of the received waveform. However as we now see, the waveform may be badly distorted.

By virtue of (2.8), the received waveform may be expressed as

$$y(t) = \text{Re}\left[\sum_i \eta_i u(t - r_i) \exp j\omega_o(t - r_i)\right] \tag{3.7a}$$

where

$$\eta_i = A\rho_i \exp -j\theta_i. \tag{3.7b}$$

We next suppose that L is much less than W^{-1} and invoke the oft-used engineering approximation that a waveform of bandwidth W does not change its value appreciably in time intervals of duration much less than W^{-1}. Thus, if

$$WL \ll 1, \tag{3.8a}$$

then

$$u(t - r) \approx u(t), \qquad \text{for} \quad |r| < \frac{L}{2}. \tag{3.8b}$$

Hence

$$y(t) \approx \text{Re} \left\{ [u(t) \exp j\omega_o t] \left(\sum_i \eta_i \exp -j\omega_o r_i \right) \right\}; \tag{3.8c}$$

that is, the received process differs from the transmitted waveform only by a scale factor (or fade) and a carrier phase shift

The approximate description for $y(t)$ provided by (3.8c) is similar in form to the exact description (3.4b) obtained under the assumption that the channel is nondispersive. The received waveform in both situations differs from the transmitted waveform only through a random fixed scale factor and a random fixed carrier phase shift. Moreover the Gaussian assumption ensures that the statistical properties of the scale factor and phase in the two situations are identical. Thus a channel that is dispersive only in time behaves as though it is nondispersive if the waveform bandwidth W is much less than L^{-1}.

On the other hand, as W is increased, a point is eventually reached (when WL is approximately unity) at which (3.8b) is violated for some values of r; that is, the returns from some of the scatterers combine destructively, or interfere. This interference occurs even though the duration T of $u(t)$ remains much greater than L. Thus, if L is much less than T but much greater than W^{-1}, there is a negligible amount of time spreading apparent in the received waveform, but the structure of the received and transmitted waveforms may differ drastically as illustrated in Fig. 3.6a. Figure 3.6b illustrates the nature of the received waveform when L exceeds T.

In summary, one manifestation of time dispersion in a channel is the presence of time spreading or distortion or both in the received waveforms. The severity of these effects depends upon both the nature of the transmitted waveform and the channel scattering function. However to a first approximation, they may be classified according to the relative values of T, W, and L as is done in Table 3.3.† This classification is fairly valid if $u(t)$ is concentrated

---

† Here and elsewhere we use the fact that the TW product of a waveform cannot be too much less than one [12, 13].

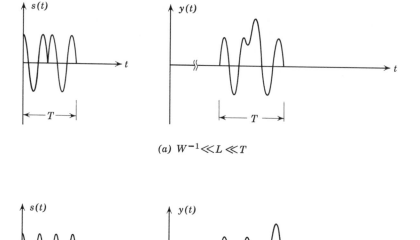

(a) $W^{-1} \ll L \ll T$

(b) $T \ll L$

Fig. 3.6 Representative waveforms for channels that are dispersive only in time; $s(t)$ transmitted, $y(t)$ received.

in time and frequency and if, in addition, the scattering function is concentrated in range delay.

Table 3.3 Waveform Characteristics for Channels Dispersive only in Time

WL	L/T	Distorted	Dispersed	Remarks
$\ll 1$	$\ll 1$	No	No	Implied by $WL \ll 1$ since $TW \geq 1$
$\gg 1$	$\ll 1$	Yes	No	Implies $TW \gg 1$
$\gg 1$	$\gg 1$	Yes	Yes	Implied by $L \gg T$ since $TW \geq 1$

Note that the situation described by the second entry of Table 3.3 arises only when the TW product of $u(t)$ is much greater than unity. In fact, if TW is unity, WL equals L/T, and the conditions $WL \gg 1$, $L/T \ll 1$ are incompatible. Thus, for unity TW products, the received waveform is spread and distorted if WL is much greater than unity, whereas it is neither spread nor distorted if

WL is much less than unity. More generally, regardless of the *TW* product, the received waveform is undistorted if *WL* is much less than unity; whereas it is distorted if *WL* is much greater than unity. However the time spreading of the waveform is determined by L/T rather than by *WL*.

The conclusions of the preceding paragraphs can also be deduced from the Fourier transforms of the transmitted and received waveforms. The Fourier transform $S(f)$ of the transmitted waveform is

$$S(f) = \tfrac{1}{2}[U(f - f_o) + U^*(-f - f_o)], \tag{3.9}$$

where $U(f)$ is, as before, the Fourier transform of $u(t)$. By virtue of (3.7), the Fourier transform $Y(f)$ of the received waveform is

$$Y(f) = \tfrac{1}{2}U(f - f_o)H(f) + \tfrac{1}{2}U^*(-f - f_o)H^*(-f), \tag{3.10a}$$

where

$$H(f) = \sum_i \eta_i \exp -j2\pi f r_i. \tag{3.10b}$$

Equations 3.9 and 3.10 display the effect of time dispersion as the attenuation of certain frequency components of the transmitted waveform. This interpretation provides a more precise description of the distortion introduced by the constructive and destructive combination of the contributions from different scatterers. Because of this description, the manifestation of time dispersion is often called *frequency-selective fading* [14–16].

Equations 3.9 and 3.10 also imply that the received and transmitted waveforms possess the same structure if *WL* is much less than unity. In fact, the condition

$$WL \ll 1$$

implies that

$$H(f) \approx \sum_i \eta_i, \quad \text{for } |f| < W; \tag{3.11}$$

that is, $Y(f)$ differs from $S(f)$ only by a complex constant. Thus we conclude, as before, that a channel that is dispersive only in time appears to be nondispersive for a given $u(t)$ if *WL* is much less than unity. The general character of $S(f)$ and $Y(f)$ when *WL* exceeds unity is shown in Fig. 3.7.

Channels Dispersive Only in Frequency

We say that a channel is dispersive only in frequency if its scattering function is of the form

$$\sigma(r, f) = \delta(r)\sigma(f), \tag{3.12}$$

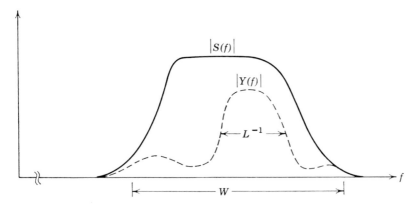

Fig. 3.7   Representative transforms for channels that are dispersive only in time; $WL > 1$; transform of transmitted waveform, $S(f)$; transform of received waveform, $Y(f)$.

where $\sigma(f)$ is the frequency scattering function of (3.1e). In writing this expression, we take the common range delay of the scatterers to be zero.

   Channels that are dispersive only in frequency are, in many ways, the duals of channels that are dispersive only in time [17]. In particular let us consider the normalized complex envelope $z(t)$ of the output process that results from using an input envelope $u(t)$ with a scattering function $\delta(r)\sigma(f)$. According to Section 2.6, $z(t)$ is the Fourier transform of the normalized envelope that results when the inverse Fourier transform of $u(t)$ is used as an input envelope to the channel whose scattering function is $\delta(f)\sigma(r)$.† Thus if we only consider complex envelopes, the input and output of the channel with range-delay scattering function $\sigma(r)$, which is dispersive only in time, are related to each other in the same manner as the Fourier transforms of the envelopes are in a channel with frequency scattering function $\sigma(f)$, which is dispersive only in frequency. Thus the discussion of channels dispersive only in time may, for the most part, be applied to channels dispersive only in frequency: one need merely interchange the roles of $u(t)$ and its Fourier transform $U(f)$, or equivalently of T and W, and also replace L by B.

   For example, if the bandwidth W of $u(t)$ is much less than the Doppler spread B, the received waveform bandwidth is appreciably greater than W; that is, the channel spreads the transform of the transmitted waveform. On the other hand, if the Doppler spread B is much less than the reciprocal duration T^{-1} of $u(t)$, the transmitted and received waveforms differ only by a scale factor and a carrier phase shift; that is, a channel dispersive only in frequency appears to be nondispersive if BT is much less than unity.

---

† We have used the fact that $\delta(f) = \delta(-f)$.

A third situation arises when B/W is much less than unity but BT is much greater than unity. These conditions are analogous to those encountered in the discussion of channels dispersive only in time and, as mentioned there, they do not occur unless the TW product of $u(t)$ is much greater than unity. When they do occur, little or no spreading is apparent in the transform of the received waveform, but it may be severely distorted.

The conclusions of the preceding discussion are summarized in Table 3.4

Table 3.4   Waveform Characteristics for Channels Dispersive only in Frequency

BT	B/W	Distorted	Dispersed	Remarks
$\ll 1$	$\ll 1$	No	No	Implied by $BT \ll 1$ since $TW \geq 1$
$\gg 1$	$\ll 1$	Yes	No	Implies $TW \gg 1$
$\gg 1$	$\gg 1$	Yes	Yes	Implied by $B \gg W$ since $TW \geq 1$

and illustrated in Fig. 3.8. We stress again that they lose their validity when

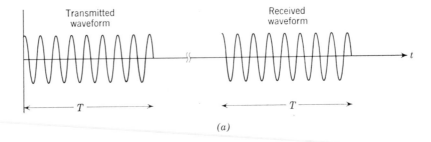

(a)

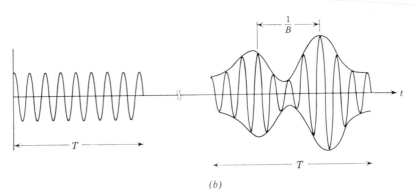

(b)

Fig. 3.8   Representative waveforms for channels that are dispersive only in frequency. (*a*) $BT < 1$, (*b*) $BT > 1$.

either $u(t)$ is not concentrated in time and frequency or when the scattering function is not concentrated in frequency.

Another description of channels that are dispersive only in frequency is afforded by (3.10) in conjunction with the duality that exists between time and frequency dispersion. In particular the complex envelope of the output process is the input envelope times a complex function. Thus

$$y(t) = \sqrt{2E_r} \, \mathrm{Re} \left[g(t)u(t) \exp j\omega_0 t \right], \qquad (3.13)$$

where $g(t)$ is a complex zero-mean Gaussian random process with

$$\overline{g(t)g(\tau)} = 0 \qquad (3.14a)$$

and

$$\overline{g(t)g^*(\tau)} = \mathcal{R}(0, \tau - t). \qquad (3.14b)$$

That is, the channel may be thought of as producing the received waveform by amplitude and phase modulating the transmitted waveform with the magnitude and the argument of $g(t)$, respectively. Viewed in this way, the channel selectively alters certain time segments of the transmitted waveform; therefore the manifestation of frequency dispersion is sometimes called *time-selective fading* [11]. The phrase *frequency-flat fading* is also employed since all the constituent frequencies of the transmitted waveform are modulated by the same function [11].

Doubly Dispersive Channels

Channels that are dispersive in both time and frequency will be called *doubly dispersive*. In general, such channels exhibit both time-selective and frequency-selective fading; stated alternately, the fading usually is neither time-flat nor frequency-flat.

The gross properties of doubly dispersive channels can be established by treating them as superpositions of channels dispersive only in frequency. That is, we first confine our attention to those scatterers that possess a given range delay r and establish the properties of their contribution to the channel output. We then determine the way in which the contributions from different range delays combine to yield the total output waveform.

The properties of the various contributions are nearly established. In fact, once we confine our attention to scatterers at a given range delay, we are left with a " channel " that is dispersive only in frequency. If the Doppler spread, say $B(r)$, of these scatterers is much greater than the bandwidth W of $u(t)$, their contribution to the output waveform is appreciably spread in frequency. On the other hand, if $B(r)$ is much less than W but much greater than T^{-1}, there is little apparent frequency spreading but severe distortion. Finally, if $B(r)$ is much less than T^{-1}, there is neither spreading nor distortion.

The validity of the above comments is, of course, predicated upon the assumption that the scatterers at the given range delay are rather smoothly distributed over a frequency interval of length $B(r)$. In order to account for the interaction of the contributions from the collections of scatterers at different range delays, we must also consider the absolute values of the Doppler shifts.

If $B(r)T$ is much less than unity for all values of r, the scatterers at any given range delay r appear to possess the same Doppler shift, say $f(r)$. Their contribution to the output waveform then may be approximately expressed as

$$\text{Re}\ \{\eta(r)u(t-r)\exp j[\omega_o(t-r)-2\pi f(r)t]\},$$

where $\eta(r)$ is the sum of the η_i (3.7b) over all those scatterers at the range delay r. The total received waveform then becomes

$$y(t) = \text{Re}\left\{\sum_r \eta(r)u(t-r)\exp j[\omega_o(t-r)-2\pi f(r)t]\right\}, \qquad (3.15)$$

where the summation is over all values of range delay at which scatterers are located.

Let us now further assume that $f(r)T$, as well as $B(r)T$, is much less than unity. This assumption often, but not always, is satisfied if BT is much less than unity. When it is satisfied, each of the exponential terms $\exp j[-2\pi f(r)t]$ essentially equals some constant $C(r)$ over the time interval for which $u(t-r)$ is nonzero, and the output waveform is given approximately by the expression

$$y(t) = \text{Re}\left[\sum_r \eta(r)C(r)u(t-r)\exp j\omega_o(t-r)\right]. \qquad (3.16)$$

That is, the received waveform appears to be the output of a channel that is dispersive only in time, and questions pertaining to time spreading and distortion can be answered as they are in Section 3.3.

As the value of T is increased, one may eventually reach a point at which either the quantities $tf(r)$ can no longer be regarded as constants or $B(r)T$ exceeds unity for some value of r. As T is further increased, distortion due to frequency dispersion begins to appear. Whether or not this distortion involves frequency spreading of the received waveform is, to a first crude approximation, determined as it is in the discussion of channels dispersive only in frequency; that is, if B/W is much less than unity, little spreading is apparent, whereas if B/W is much greater than unity, there is appreciable spreading.

After we determine the effects of frequency dispersion upon the output waveform, the effects of time dispersion can be evaluated to arrive at the gross conclusions summarized in Table 3.5. Since the procedure is identical to that applied to channels dispersive only in time, we do not repeat it here.

Table 3.5   Waveform Characteristics for Doubly Dispersive Channels

BT	WL	B/W	L/T	Distorted	Dispersed in Time	Dispersed in Frequency	Remarks
$\ll 1$	$\ll 1$	$\ll 1$	$\ll 1$	No	No	No	Implies $BL \ll 1$
$\ll 1$	$\gg 1$	$\ll 1$	$\ll 1$	Yes	No	No	Implies $BL \ll 1$, and $TW \gg 1$
$\ll 1$	$\gg 1$	$\ll 1$	$\gg 1$	Yes	Yes	No	
$\gg 1$	$\ll 1$	$\ll 1$	$\ll 1$	Yes	No	No	Implies $BL \ll 1$, and $TW \gg 1$
$\gg 1$	$\gg 1$	$\ll 1$	$\ll 1$	Yes	No	No	Implies $TW \gg 1$
$\gg 1$	$\gg 1$	$\ll 1$	$\gg 1$	Yes	Yes	No	Implies $BL \gg 1$, and $TW \gg 1$
$\gg 1$	$\ll 1$	$\gg 1$	$\ll 1$	Yes	No	Yes	
$\gg 1$	$\gg 1$	$\gg 1$	$\ll 1$	Yes	No	Yes	Implies $BL \gg 1$, and $TW \gg 1$
$\gg 1$	$\gg 1$	$\gg 1$	$\gg 1$	Yes	Yes	Yes	Implies $BL \gg 1$

However it is important to note that the validity of Table 3.5 is even more limited than is that of the corresponding tables (3.3 and 3.4) for channels dispersive only in time and only in frequency. This is so because the character of the received waveform is influenced by the shape of the scattering function, and this shape is not uniquely determined by B and L even when the scattering function is known to be "smooth" and "concentrated." These limitations are the price we must pay for the simple description we seek.

We have now related the gross character of the received waveform to B, L, T, and W. We next establish a slightly more quantitative description by examining the average power and coherence properties of the received waveform. This examination yields further insight into the significance of B, L, T, and W and enables us to establish a bridge, in Section 3.6, between the precise but complicated results of Chapter 2 and the simple but imprecise descriptions that are frequently employed.

3.4 AVERAGE POWER

The average power $\bar{P}(t)$ received at a time t is defined to be

$$\bar{P}(t) = \overline{[y(t)]^2}. \tag{3.17}$$

Recalling the definition of the correlation function $R_y(t, \tau)$, we express this power as

$$\bar{P}(t) = R_y(t, t) \tag{3.18a}$$

or

$$\bar{P}(t) = E_r \int |u(t-r)|^2 \, \sigma(r) \, dr. \qquad (3.18b)$$

If $|u(t)|^2$ is an impulse function, (3.18b) becomes

$$\bar{P}(t) = E_r \sigma(t); \qquad (3.19)$$

that is, the profile of the time distribution of average power is proportional to the range-delay scattering function of the channel. More generally, $\bar{P}(t)$ is proportional to $\sigma(t)$ if $|u(t)|^2$ "appears" to be impulsive with respect to $\sigma(t)$: a circumstance that prevails if the duration T of $u(t)$ is sufficiently less than the multipath spread L. More precisely, (3.19) is valid if the Fourier transform of $|u(t)|^2$ is constant over the frequency interval where the Fourier transform of $\sigma(r)$ is nonzero.

On the other hand, if L is sufficiently small, $\sigma(t)$ appears to be an impulse with respect to $u(t)$, and the expression for $\bar{P}(t)$ becomes

$$\bar{P}(t) = E_r |u(t)|^2. \qquad (3.20)$$

This expression is approximately valid if L is much less than the reciprocal bandwidth of $|u(t)|^2$; it is precisely valid if the Fourier transform of $\sigma(t)$ equals a constant for all values of f for which the Fourier transform of $|u(t)|^2$ is nonzero.

A third situation arises if L is much less than T but greater than the reciprocal bandwidth of $|u(t)|^2$. Then, the simplifications that yield (3.19) and (3.20) are invalid, and all we can say is that $\bar{P}(t)$ is positive over a time interval of approximately $T + L$ sec. This situation arises only if the (real) envelope $|u(t)|$ of the transmitted waveform has a TW product that exceeds unity; it cannot arise if the transmitted waveform possesses a unit TW product or if a large TW product is achieved by phase or frequency modulation. For waveforms whose (real) envelopes possess time-bandwidth products of unity, we may summarize our results as follows: The average power is given approximately by (3.19) if L/T is much greater than unity and by (3.20) if L/T is much less than unity.

It is also useful to consider the distribution of average power in the frequency domain. Thus we consider the average energy density of the normalized complex envelope of the received waveform. We focus upon the complex envelope rather than upon the waveform itself, so as to avoid some bothersome and unimportant details involving the carrier frequency.

The density we seek follows easily from (2.17). It is

$$\overline{|Z(f)|^2} = \int \sigma(-\tilde{f}) |U(f-\tilde{f})|^2 \, d\tilde{f}, \qquad (3.21)$$

where $Z(f)$ is the Fourier transform of $z(t)$ and, as before, $U(f)$ is the Fourier transform of $u(t)$. This expression is identical in form to the expression (3.18b) for the average power distribution of the received waveform. Aside from unimportant scale factors, they differ only in the substitution of $\sigma(-f)$ for $\sigma(r)$ and the substitution of $|U(f)|^2$ for $|u(t)|^2$. Consequently we may deduce the properties of $\overline{|Z(f)|^2}$ from our knowledge of $\overline{P(t)}$. For example, the average energy density of the received waveform is proportional to $\sigma(-f)$ when $U(f)$ is an impulse, that is, when an unmodulated carrier is transmitted. More generally, $\overline{|Z(f)|^2}$ is positive over a frequency interval of about $B + W$ Hz.

3.5 WAVEFORM COHERENCE

Our objective, in this section, is to estimate the statistical dependencies between different portions of the received waveform in both the time and the frequency domains. Precisely stated, we seek an estimate of the time separation T_c beyond which samples of the complex envelope of the received process are independent. Similarly we seek an estimate of the frequency separation W_c beyond which samples of the Fourier transform of the complex envelope are independent. These separations are called, respectively, the *correlation time* and *correlation bandwidth* of the process. The terms *coherence time* and *coherence bandwidth* are also frequently used.

We employ the complex envelope of the process rather than the process itself because our real interest is in the intervals beyond which samples of the low-pass quadrature components of the process are independent. That is, we are not concerned with the independence that can be achieved with samples separated by time intervals of less than one period of the carrier.

Coherence Time

Since the received process and its complex envelope are Gaussian and of mean zero, the requirement that two samples be independent is equivalent to the requirement that they be uncorrelated. Thus we inquire about the difference $t - \tau$ for which the complex correlation function $R(t, \tau)$ vanishes. By virtue of (2.22b), (2.25a), and (2.26b), this condition may be stated as

$$R(t, \tau) = 0 \qquad (3.22a)$$

where

$$R(t, \tau) = \int \mathscr{R}(\alpha, \tau - t)\theta(\tau - t, \alpha)\, \exp\, -j\pi\alpha(t + \tau)\, d\alpha, \qquad (3.22b)$$

$$\mathscr{R}(\alpha, \tau - t) = \iint \sigma(r, f)\, \exp j2\pi[r\alpha + f(\tau - t)]\, dr\, df, \qquad (3.22c)$$

and

$$\theta(\tau - t, \alpha) = \int u\left(x - \frac{\tau - t}{2}\right)u^*\left(x + \frac{\tau - t}{2}\right)\exp j2\pi\alpha x \, dx. \qquad (3.22d)$$

For channels that are dispersive only in frequency, the right member of (3.22b) becomes $u(t)\,\mathcal{R}(0, \tau - t)u^*(\tau)$ so that (2.22a) is satisfied if any of the following conditions are met:

$$u(t) = 0, \qquad\qquad\qquad (3.23a)$$

$$u(\tau) = 0, \qquad\qquad\qquad (3.23b)$$

$$\mathcal{R}(0, \tau - t) = 0. \qquad\qquad\qquad (3.23c)$$

The first two of these equations imply that the average power in one or both of the samples is zero. Since samples of zero power are of no interest to us, (2.23c) determines the time separation T_c beyond which two samples are independent. Specifically T_c is the smallest time for which

$$\mathcal{R}(0, x) = 0, \qquad \text{for all} \quad |x| > T_c. \qquad (3.24)$$

If $\mathcal{R}(0, x)$ is a fairly smooth function of x, its half width does not differ too much from T_c. Moreover this half width is roughly equal to B^{-1}, the reciprocal of the width of the frequency scattering function.† That is, T_c is very roughly equal to B^{-1} where B is the Doppler spread of the channel.

The value of T_c for doubly dispersive channels depends upon the waveform employed. For example, if the modulating bandwidth W is much less than L^{-1}, the channel appears to be only frequency dispersive (Table 3.5) and hence the correlation constant T_c is approximately B^{-1}. More generally, by virtue of (3.22), T_c may be taken to be any number for which

$$\mathcal{R}(\alpha, t)\theta(t, \alpha) \approx 0 \qquad\qquad (3.25)$$

for all values of α and for all values of t greater than T_c.

We can derive a variety of crude upper bounds to the value of T_c that satisfies (3.25). One particularly simple and useful bound is

$$T_c \leq T, \qquad\qquad (3.26)$$

where T is the duration of the transmitted waveform. Its validity may be established by noting either that $\theta(T, \alpha) \approx 0$ or that any given scatterer influences the received process for a time interval of T sec. Therefore the behavior of the process at any two points separated by more than T sec is determined by different sets of scatterers. Since the cross sections and phases

---

† This is precisely true for a rectangular scattering function (the first entry of Table 3.2). For other scattering functions T_c may differ from B^{-1}.

of different scatterers are presumed to be statistically independent, so also is the behavior of the received process at the two times in question.

At this juncture we have found that, to a first approximation, T_c is less than both T and B^{-1}. It is interesting to observe that T_c may be appreciably less than both T and B^{-1}. To illustrate this possibility, we consider the time-dispersive channel whose scattering function is

$$\sigma(r) = \frac{\sqrt{2}}{L} \exp - \frac{2\pi r^2}{L^2} \qquad (3.27a)$$

and suppose that the transmitted waveform is a "chirped" Gaussian pulse [18], that is,

$$u(t) = \left(\frac{\sqrt{2}}{T}\right)^{1/2} \exp - \pi \left(\frac{t}{T}\right)^2 [1 + j\sqrt{(TW)^2 - 1}]. \qquad (3.27b)$$

The introduction of the above expressions into (3.18) and (3.22) yields

$$\bar{P}(t) = \frac{\sqrt{2E_r}}{\sqrt{L^2 + T^2}} \exp - \frac{2\pi t^2}{L^2 + T^2} \qquad (3.28a)$$

and

$$R(t, \tau) = \left(\frac{2}{L^2 + T^2}\right)^{1/2} \left[\exp - \frac{\pi}{2} \frac{(LW)^2}{L^2 + T^2} (t - \tau)^2\right]$$

$$\times \exp - \frac{\pi}{T^2 + L^2} [t^2(1 + j\beta) + \tau^2(1 - j\beta)], \qquad (3.28b)$$

where

$$\beta = \sqrt{(TW)^2 - 1}. \qquad (3.28c)$$

Finally we suppose that T is much greater than L, so that

$$R(t, \tau) \approx 0, \qquad \text{for} \quad t - \tau \ge \frac{T}{WL}. \qquad (3.29)$$

Thus the coherence time T_c is approximately equal to T/WL. Clearly T_c is less than both B^{-1} and T if W is sufficiently large.

Coherence Bandwidth

The coherence bandwidth of the received process is the frequency separation beyond which samples of the Fourier transform of the complex envelope are independent. By virtue of the equivalence discussed in Section 2.6 and employed in Section 3.5 this transform behaves in every respect as though it is the complex envelope resulting from the transmission of $U(t)$ [rather than $u(t)$] through the channel whose scattering function is $\sigma(f, -r)$ [rather than $\sigma(r, f)$]. Thus the coherence bandwidth W_c exceeds neither L^{-1} nor W.

We have now completed our gross characterization of the received waveform in terms of B, L, T, and W. Although these parameters do not replace the diversity description of Section 2.5, they are often useful in estimating the parameters of that description preparatory to a more refined analysis. We turn now to a discussion of such estimates.

3.6   DIVERSITY ESTIMATES

In general an infinite number of eigenvalues, or diversity paths, appear in Fig. 2.4. However only a finite number, say D, of these eigenvalues are large enough to be "important." Moreover in many problems of interest, the D "important" eigenvalues are roughly equal to $1/D$. Stated alternately, the Karhunen-Loève series of (2.46c) for the (complex) channel-output process consists, in essence, of D terms—the variance of each term being $1/D$. In such problems it is appropriate to regard D as the number of "effective" diversity paths, or eigenvalues, associated with the system.

Problems involving D equal eigenvalues are important because, as shown in Chapter 5, they are closely related to optimum communication systems. Moreover even when the eigenvalues are not equal, a rough simple estimate of the system performance often can be obtained by assuming that the system behaves as though its important eigenvalues are equal. Such estimates are useful because it is often laborious to evaluate the system performance when the eigenvalues are not equal. Therefore we now relate D, the number of "important" eigenvalues, to B, L, T, and W. Specifically we first upper bound and then lower bound D. Finally we introduce a precise expression for D when all of the positive eigenvalues are equal to each other. Taken together these results provide some useful estimates for D and also illustrate the shortcomings of such estimates.

Upper Bounds

We approximate the complex envelope of the received process by expressions of the form

$$z(t) = \sum_{i=1}^{N_s} a_i V_i(t), \tag{3.30}$$

where the $V_i(t)$ are a set of known functions, the a_i are Gaussian random variables, and N_s is a function of the gross system parameters.

The approximate representation of (3.30) is relevant to our discussion because it implies that the complex correlation function of the received process can be expressed as

$$R(t, \tau) = \sum_{i=1}^{N_S} \sum_{j=1}^{N_S} \overline{a_i a_j^*} V_i(t) V_j^*(\tau). \tag{3.31}$$

Correlation functions of this form are called degenerate kernels and are known to possess at most N_s positive eigenvalues [19, 20]. Thus to the extent that (3.30) is valid, N_s provides an upper bound to D. One useful form of (3.30) is obtained as follows.

We recall from (2.8) and (2.17) that the complex envelope of the received process can be expressed as

$$z(t) = \frac{1}{\sqrt{2E_r}} \sum_i \eta_i u(t - r_i) \exp -j(\omega_o r_i + \omega_i t), \tag{3.32a}$$

where

$$\eta_i = A\rho_i \exp -j\theta_i. \tag{3.32b}$$

We again denote the bandwidth of $u(t)$ by W and invoke the approximation that a waveform of bandwidth W cannot change its value appreciably in a time interval of duration W^{-1}; that is,

$$u(t - r_i) \approx u\left(t - \frac{k}{W}\right) \tag{3.33a}$$

provided that

$$\left| r_i - \frac{k}{W} \right| \leq \frac{1}{2W}. \tag{3.33b}$$

We next divide the range of summation in (3.32a) into the set of intervals

$$\left| \frac{k}{W} - r_i \right| \leq \frac{1}{2W}, \qquad k = 0, \pm 1, \pm 2, \ldots, \tag{3.34a}$$

to obtain

$$z(t) \approx \frac{1}{\sqrt{2E_r}} \sum_k \sum_{A_k} \eta_i u\left(t - \frac{k}{W}\right) \exp -j(\omega_o r_i + \omega_i t), \tag{3.34b}$$

where A_k denotes the set of those i for which (3.34a) is satisfied.

Upon defining

$$a_k(t) = \frac{1}{\sqrt{2E_r}} \sum_{A_k} \eta_i \exp -j(\omega_o r_i + \omega_i t), \tag{3.35a}$$

we obtain

$$z(t) \approx \sum_{k=-\infty}^{\infty} a_k(t) u\left(t - \frac{k}{W}\right). \tag{3.35b}$$

For simplicity we again assume that the scattering function is zero if the magnitude of r exceeds $L/2$; that is,

$$\eta_i = 0, \quad \text{if} \quad |r_i| > \frac{L}{2}.$$

Consequently

$$a_k(t) = 0, \quad \text{if} \quad |k| \geq \frac{1 + WL}{2},$$

and we may restate (3.35b) as

$$z(t) = \sum a_k(t) u\left(t - \frac{k}{W}\right), \tag{3.36}$$

where the summation is over all k such that

$$2|k| < 1 + LW.$$

Equation 3.36 expresses the received process as a combination of approximately $1 + WL$ known time functions. The (complex) functions $a_k(t)$ that multiply these known functions are statistically independent because the values of different functions are determined by different scatterers. Consequently, to the extent that we believe the engineering approximations employed, the detailed character of the received process is determined by a set of $1 + WL$ statistically independent random functions of time.

To cast (3.36) in the form of (3.30), we need only apply the (complex) sampling theorem to each of the functions $a_k(t)$ [21]. Specifically we express $a_k(t)$ as

$$a_k(t) = \sum_i a_{ik}\left[\exp j2\pi\Omega_k\left(t - \frac{i}{B_k}\right)\right] \frac{\sin \pi(i - B_k t)}{\pi(i - B_k t)}, \tag{3.37}$$

where a_{ik} is the value of $a_k(t)$ at time i/B_k and where B_k and Ω_k are, respectively, the Doppler spread and average Doppler shift of those scatters whose range delay is k/W. We then introduce (3.37) into (3.36) to obtain

$$z(t) = \sum_k \sum_i a_{ik} u\left(t - \frac{k}{W}\right)\left[\exp j2\pi\Omega_k\left(t - \frac{i}{B_k}\right)\right] \frac{\sin \pi(i - B_k t)}{\pi(i - B_k t)}, \tag{3.38}$$

where k ranges over the integers for which

$$2|k| < 1 + LW.$$

Except for the use of a double, rather than a single, index, (3.38) is of the same form as (3.30). Thus the number of terms in (3.38) provides a value for N_s. To estimate that number, we suppose that $u(t)$ vanishes outside an interval of length T. This supposition implies that, for each k, only about $1 + B_k T$ terms contribute to the summation over i. Since k ranges over the integers of magnitude less than about $(1 + WL)/2$, the total number of non-zero terms appearing in (3.38), and hence a value of N_s, is approximately

$$N_s \approx \sum (1 + B_k T), \tag{3.39}$$

where the summation is over the integers k for which

$$2|k| < 1 + LW.$$

Equation 3.39 reduces to a more familiar expression if the scattering function is approximately of the form shown in Fig. 3.9. Then B_k equals B, the Doppler spread of the channel, and

$$N_s \approx (1 + BT)(1 + LW). \tag{3.40a}$$

Somewhat more generally, if the scattering function is of the form shown in Fig. 3.10, (3.39) becomes

$$N_s \approx \left(1 + \frac{ST}{L}\right)(1 + LW). \tag{3.40b}$$

Equation 3.40a provides a reasonable estimate to the number of (complex) "degrees of freedom" associated with the channel [22, 23]. However it must

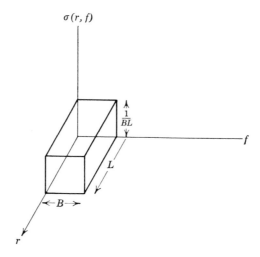

Fig. 3.9 Rectangular scattering function.

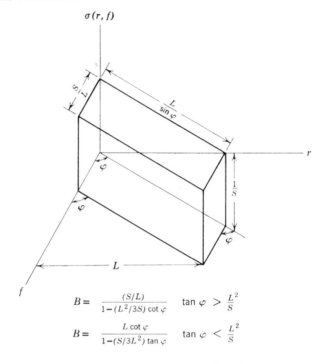

$$B = \frac{(S/L)}{1-(L^2/3S)\cot\varphi} \qquad \tan\varphi > \frac{L^2}{S}$$

$$B = \frac{L\cot\varphi}{1-(S/3L^2)\tan\varphi} \qquad \tan\varphi < \frac{L^2}{S}$$

Fig. 3.10   Skewed scattering function.

be emphasized that smaller values of N_s exist when the functions appearing in (3.38) are linearly dependent. To illustrate this fact, we apply the complex sampling theorem directly to the received waveform.

Specifically let us again suppose that the scattering function is of the form shown in Fig. 3.9. The bandwidth of the received waveform is then approximately $W + B$ Hz, and its duration is approximately $L + T$ sec. Thus the sampled expansion for the process consists of approximately $(B + W)(L + T)$ samples spaced $(B + W)^{-1}$ sec apart over a time interval of $L + T$ sec; that is, the process can be expressed in the form of (3.30) with

$$N_s = (B + W)(L + T). \tag{3.41}$$

A comparison of (3.40a) with (3.41) shows that the latter expression provides the sharper bound to D when TW and BL both exceed unity.† Therefore

$$\begin{aligned} D &\le (L + T)(B + W), &&\text{if } BL \text{ and } TW > 1, \\ D &\le (1 + BT)(1 + LW), &&\text{otherwise.} \end{aligned} \tag{3.42}$$

† The validity of this statement rests in part upon the fact that $TW \ge 1$.

There are several other avenues that can be followed to obtain better estimates of D, the number of important positive eigenvalues. For example, D is approximately equal to the rank of the covariance matrix of the samples discussed in conjunction with (3.41) [19, 20]. However, since these estimates are rather complex, we follow another avenue and supplement our upper bounds with some lower bounds to D.

Lower Bounds

We lower bound the number of statistically independent samples that can be extracted from the complex envelope of the received process. This number is important because it provides a lower bound to the number of positive eigenvalues, or diversity paths. That is, if N_i statistically independent samples can be extracted from the process, the Karhunen-Loève series for the envelope contains at least N_i terms. Thus N_i is a lower bound to D.

We first recall, from Section 3.4, that the received power is significant over a time interval of approximately $T + L$ sec. We further recall that samples of the complex envelope which are separated either by about T sec or by about B^{-1} sec are approximately statistically independent of each other. Thus we can observe the complex envelope at either $(L + T)/T$ or $(L + T)B$ instants of time and obtain a set of pairwise statistically independent samples. Moreover since the process is Gaussian, the pairwise independence of the samples implies that they are, in fact, statistically independent. Thus

$$D \geq 1 + \frac{L}{T}, \qquad \text{if } BT < 1, \qquad (3.43a)$$

$$D \geq \left(1 + \frac{L}{T}\right)BT, \qquad \text{if } BT \geq 1. \qquad (3.43b)$$

Of course, the correlation time may be significantly less than both T and B^{-1}, and hence the bounds of (3.43) may be unnecessarily weak. To illustrate this possibility, we reconsider the sample of (3.27) through (3.29). According to (3.29), samples spaced T/WL sec apart are independent when T is much greater than L. Therefore

$$D \geq (L + T) \div \left(\frac{T}{WL}\right) \approx WL.$$

When both T and W are much greater than L, this bound is much sharper than that provided by (3.43).

Another lower bound to D is obtained by considering samples of the Fourier transform of the complex envelope of the received process. According to Section 3.4, the Fourier transform possesses a bandwidth of about

$W + B$ Hz. Moreover the values of this transform at two frequencies are approximately statistically independent of each other if the frequency spacing exceeds either W or L^{-1}. Thus the same reasoning that led to (3.43) now leads us to conclude that

$$D \geq 1 + \frac{B}{W}, \qquad \text{if} \quad WL < 1, \qquad (3.44a)$$

$$D \geq \left(1 + \frac{B}{W}\right)WL, \qquad \text{if} \quad WL \geq 1. \qquad (3.44b)$$

The upper and lower bounds to D differ by at most a factor of 2 when the scattering function is of the form shown in Fig. 3.9 and $TW = 1$. However when the TW product becomes large, the bounds may differ by a factor of at least TW for the simple scattering function of Fig. 3.9, and for more general scattering functions the discrepancy can be even worse. Thus these bounds must be used with discretion. Of course, the results can be refined at the cost of additional complexity, but a question still remains of whether or not the "important" eigenvalues are equal. Since our original plan was to use D as an estimate of the common value of the positive eigenvalues we now shift our approach and introduce an expression for D under the premise that the positive eigenvalues are all equal.

Equal Eigenvalues

Let us suppose that each of the D positive eigenvalues equals D^{-1}. Then

$$D = \frac{1}{b}, \qquad (3.45)$$

where

$$b = \sum_i \lambda_i^2. \qquad (3.46a)$$

Or, by virtue of (2.42),

$$b = \iint |R(t, \tau)|^2 \, dt \, d\tau. \qquad (3.46b)$$

We next introduce (2.26b) into (3.46b) and interchange the order of integration to obtain

$$b = \iint |\mathscr{R}(x, y)\theta(y, x)|^2 \, dx \, dy, \qquad (3.47)$$

where $\mathscr{R}(x, y)$ is the two-frequency correlation function of the channel (2.22) and $\theta(y, x)$ is the two-dimensional correlation function of the complex

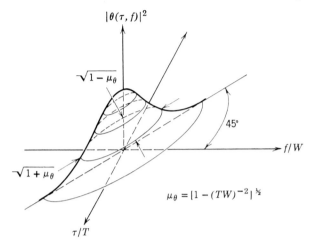

Fig. 3.11 $|\theta(\tau, f)|^2$ for a chirped Gaussian pulse. Cross sections are ellipses with major axis at 45° from f/W axis. Ratio of lengths of major and minor axis is $[(1 + \mu_\theta)/(1 - \mu_\theta)]^{1/2}$.

envelope $u(t)$ (2.25). The quantity $|\theta(y, x)|^2$ is the ambiguity function encountered in radar problems [24–26].

In a few instances (3.47) can be expressed in closed form. For example, it is shown in Chapter 6 (6.31) that

$$b = \left[1 + S^2 + (BT)^2 + (LW)^2 - 2\sqrt{(TW)^2 - 1}\,\sqrt{(BL)^2 - S^2}\right]^{-1/2} \quad (3.48)$$

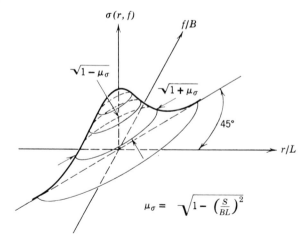

Fig. 3.12 The doubly Gaussian-shaped scattering function. Cross sections are ellipses with major axis at 45° from r/L axis. Ratio of lengths of major and minor axis is $[(1 + \mu_\sigma)/(1 - \mu_\sigma)]^{1/2}$.

when the scattering function is

$$\sigma(r, f) = \frac{2}{S} \exp - \frac{2\pi}{S^2} \left[(rB)^2 + (fL)^2 - 2rf\sqrt{(BL)^2 - S^2} \right] \qquad (3.49a)$$

and the complex modulation envelope is

$$u(t) = \frac{2^{1/4}}{\sqrt{T}} \exp - \pi \left(\frac{t}{T} \right)^2 \left[1 + j\sqrt{(TW)^2 - 1} \right] \qquad (3.49b)$$

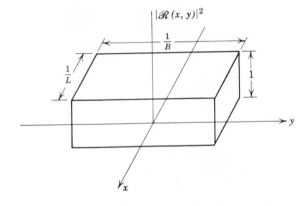

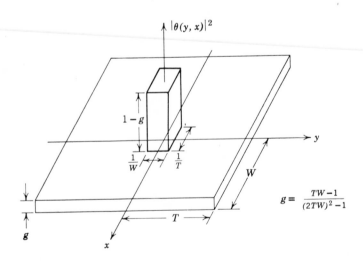

Fig. 3.13 Idealized forms of $|\mathscr{R}(\alpha, \beta)|^2$ and $|\theta(\tau, f)|^2$ for diversity estimates.

The forms of $|\theta(\tau, f)|^2$ and $\sigma(r, f)$ for this scattering function and complex envelope are shown in Figs. 3.11 and 3.12.

As one might expect, there is no simple general correspondence between (3.48) and our previous bounds. However when BL equals S and TW equals one, the results obtained from (3.48) and (3.42) are in reasonable agreement. Thus our faith in those bounds is enhanced—at least when the TW product of the complex envelope equals one and the scattering function is approximately of the form shown in Fig. 3.9, that is, when it is concentrated in time and frequency and is not skewed.

When (3.47) cannot be expressed in closed form, its value can still sometimes be estimated in a relatively simple way. To illustrate the technique, we suppose that $|\mathcal{R}(\alpha, \beta)|^2$ and $|\theta(\tau, f)|^2$ can be reasonably approximated by functions of the form shown in Fig. 3.13. That is, the two-frequency correlation function is concentrated in time and frequency and is unskewed, and the ambiguity function is the idealized "thumbtack" [24–26]. The role of the parameters B, L, T, W, and S in these figures follows from the properties of ambiguity functions and the definitions of (3.1). To the extent that we accept the approximations of Figs. 3.13, (3.47) can be evaluated as suggested by Fig. 3.14 to obtain the results summarized in Table 3.6.

Table 3.6 and (3.42) yield values of D that differ by at most a factor of 2 when TW is large. Thus we are again encouraged to use (3.42) as an estimate of D for scattering functions that are "concentrated" and unskewed (Fig. 3.9). However it is important to realize that this estimate can be seriously in error when the scattering function consists of several disjoint pieces.

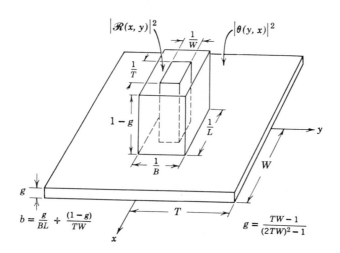

Fig. 3.14 Graphical estimation of D; $1/2W < L < T$; $1/2T < B < W$.

One means of treating such scattering functions is outlined in Section 6.1.

Table 3.6   Values of b, Equal D^{-1}, Estimated from the Construction of Fig. 3.14.

$$g = \frac{TW-1}{(2TW)^2 - 1}$$

B			
W	$b = \dfrac{1}{BT} + \dfrac{g}{B}\left(2W - \dfrac{1}{T}\right)$	$b = \dfrac{1}{BT} + \dfrac{g}{B}\left(\dfrac{1}{L} - \dfrac{1}{T}\right)$	$b = \dfrac{1}{BL}$
	$b = \dfrac{1}{TW} + g\left(\dfrac{2W}{B} - \dfrac{1}{TW}\right)$	$b = \dfrac{1}{TW} + g\left(\dfrac{1}{BL} - \dfrac{1}{TW}\right)$	$b = \dfrac{1}{LW} + \dfrac{g}{L}\left(\dfrac{1}{B} - \dfrac{1}{W}\right)$
$1/2T$	$b = 1$	$b = \dfrac{1}{TW} + \dfrac{g}{T}\left(\dfrac{2T^2}{L} - \dfrac{1}{W}\right)$	$b = \dfrac{1}{LW} + \dfrac{g}{L}\left(2T - \dfrac{1}{W}\right)$
	$1/2W$		T → L

REFERENCES

[1] S. Silver, Ed., *Monograph on Radio Waves and Circuits*. P. E. Green, Jr., "Time-Varying Channels with Delay Spread," New York: Elsevier, 1963, pp. 208.

[2] J. V. Evans and T. Hagfors, Eds., *Radar Astronomy*. P. E. Green, Jr., New York: McGraw-Hill, 1968, Chapter 1, pp. 33.

[3] G. H. Pettengill, Lecture 10, "Lunar Studies in Radar Astronomy." M.I.T. Summer Program, Aug, 14–18, 1961.

[4] I. L. Lebow, P. R. Drouilhet, N. L. Daggett, J. N. Harris, and F. Nagy, "The West Ford Belt as a Communications Medium." *Proc. IEEE*, pp. 543–563, May 1964.

[5] R. M. Lerner, "Means for Counting Effective Numbers of Objects or Durations of Signals." *Proc. IRE*, **47**, p. 1653, September 1959.

[6] J. M. Wozencraft and I. M. Jacobs, *Principles of Communication Engineering*. New York: Wiley, 1965, pp. 532.

[7] J. N. Pierce, "Theoretical Diversity Improvement in Frequency-Shift Keying." *Proc. IRE*, pp. 903–910, May 1958.

[8] G. L. Turin, "Error Probabilities for Binary Symmetric Ideal Reception through Nonselective Slow Fading and Noise." *Proc. IRE*, pp. 1603–1619, September 1958.

[9] D. G. Brennan, "Linear Diversity Combining Techniques." *Proc. IRE*, pp. 1075–1102, June 1959.

[10] H. B. Voelcker, "Phase-Shift Keying in Fading Channels." *Proc. Inst. Elec. Engrs.*, part B, pp. 31–38, January 1960.

[11] P. A. Bellow and B. D. Nelin, "The Effect of Frequency Selective Fading on the Binary Error Probabilities of Incoherent and Differentially Coherent Matched Filter Receivers." *IEEE Trans. Commun. Systems*, pp. 170–186, June 1963.

[12] J. M. Wozencraft and I. M. Jacobs, *Principles of Communication Engineering*. New York:Wiley, 1965, 294–297.

[13] H. L. Landau and H. P. Pollack, "Prolate Spheroidal Wave Functions, Fourier Analysis and Uncertainty-III: The Dimension of the Space of Essentially Time and Band-Limited Signals." *Bell System Tech. J.*, 1295–1336, July 1962.

[14] U.S. Department of Commerce, *Ionospheric Radio Propagation*, pp. 253, 1965.

[15] M. Schwartz, W. R. Bennett, and S. Stein, *Communication Systems and Techniques*, New York: McGraw-Hill, 1966, pp. 351.

[16] J.T.A.C., "Radio Transmission by Ionospheric and Tropospheric Scatter." *Proc. IRE*, pp. 4–29, January 1960.

[17] P. A. Bello. "Time-Frequency Duality." *IEEE Trans. Inform. Theory*, 18–33, January 1964.

[18] P. M. Woodward, *Probability and Information Theory, with Applications to Radar*. London: Pergamon, 1953, 123, 124.

[19] R. Courant and D. Hilbert, *Methods of Mathematical Physics*. New York: Interscience, 1953, Chapter 3.

[20] F. G. Tricomi, *Integral Equations*. New York: Interscience, 1957.

[21] P. M. Woodward, *Probability and Information Theory, with Applications to Radar*. London: Pergamon, 1953, pp. 31–35.

[22] T. Kailath, "Measurements on Time-Variant Communication Channels." *IEEE Trans. Inform. Theory*, pp. 229–236, September 1962.

[23] R. S. Kennedy and I. L. Lebow, "Signal Design for Dispersive Channels." *IEEE Spectrum*, pp. 231–237, March 1964.

[24] P. M. Woodward, *Probability and Information Theory, with Applications to Radar*. London: Pergamon, 1953, Chapter 7.

[25] J. V. Evans and T. Hagfors, Eds., *Radar Astronomy*. P. E. Green, New York: McGraw-Hill, 1968, Chapter 1, pp. 35–37.

[26] W. W. Harman, *Principles of the Statistical Theory of Communication*. New York: McGraw-Hill, 1963, pp. 279–284.

Chapter 4

Communication Systems

The system aspects of fading dispersive communication systems are the subject of this chapter. Our objectives are to describe the system that is analyzed in subsequent chapters and to reduce that description to a form suitable for analysis. We begin by considering the communication system depicted in Figs. 4.1 through 4.3. Then, by imposing constraints upon the waveforms employed, we reduce that system to the one shown in Figs. 4.5 and 4.9. The constraints are, in essence, that there be no intersymbol interference or memory and that orthogonal waveforms be employed, for example, *m*-ary frequency shift keying. We also develop one form of the optimum receiver (Fig. 4.7) and describe some techniques that can be used to satisfy the constraints we impose.

4.1 SYSTEM STRUCTURE

In this and the following chapters we are concerned with communication systems of the form shown in Fig. 4.1. As indicated, the input to the system is a sequence of binary information symbols occurring at a rate of R symbols, or bits, per second. The only real restriction is that the symbol alphabet be finite; the supposition that it is binary causes no additional loss of generality. For simplicity we will suppose that the two binary symbols occur with equal probability and that successive symbols are statistically independent of each other. These suppositions imply that the information rate of the input sequence is R bits/sec.

The output of the system also consists of a stream of binary symbols occurring at a rate of R bits/sec. The objective is to obtain an output sequence that reproduces the input sequence to within some specified degree of fidelity. For our purposes, it is convenient to measure this fidelity, or performance, by a suitably defined error probability. More is said about this later.

The transmitter converts the sequence of information bits into a sequence of waveforms which propagates through the channel and is added to zero-mean white Gaussian noise to yield the received process $r(t)$. The power

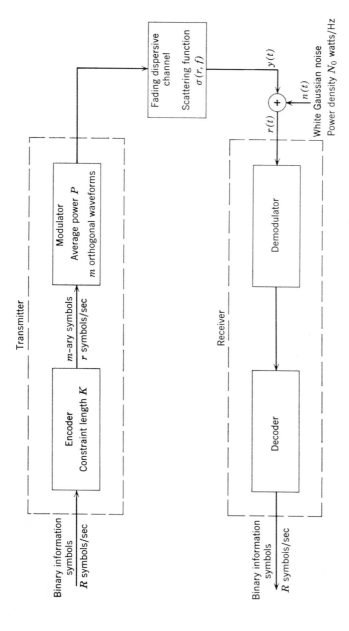

Fig. 4.1 Block diagram of fading dispersive communication system.

density of the noise is denoted by N_0 watts/Hz based upon a single-sided (or measured) spectrum; that is, the bilateral power-density spectrum has value $N_0/2$ for all values of f. The assumption that the noise is white is made primarily for convenience: it is only necessary that the noise power-density spectrum be essentially flat over the frequency band of the channel-output waveform.

The complete communication system is often most conveniently designed and analyzed as a set of subsystems composed of the encoder and decoder on the one hand and of the modulator and demodulator on the other. These subsystems differ primarily in the size of the symbol alphabets they employ and in the time interval over which they operate.

In subsequent chapters we emphasize systems that do not employ coding. We do this because the available techniques and results of coding theory can be applied to the system of Fig. 4.1 with relative ease once the characteristics of the (uncoded) modulator-demodulator system are understood [1–4]. However in this chapter we describe both aspects of the system so as to facilitate the application of coding results. This application is illustrated in Chapter 5.

Waveform Generation

As illustrated in Figs. 4.1 and 4.2, we envisage the transmitter as being composed of an encoder and a modulator.

The encoder accepts the binary information sequence as its input and generates, as its output, a sequence of m-ary symbols occurring at a rate of r symbols/sec. The encoder may be envisaged as the concatenation of two devices: a binary-to-binary encoder followed by a binary–to–m-ary converter (Fig. 4.3). When m^r equals 2^R, that is, when

$$r = \frac{R}{\log_2 m},\qquad(4.1)$$

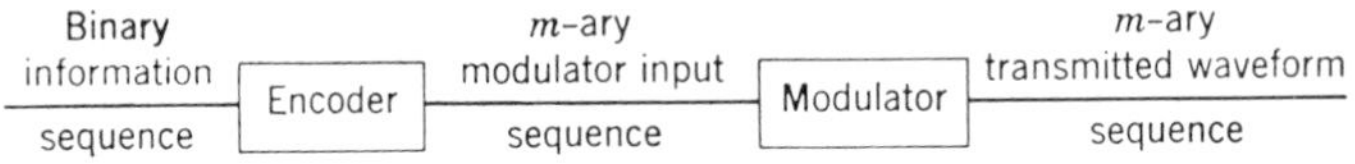

Sequences:

Information	$\xi_1, \xi_2, \ldots$	$\xi_i = (0, 1)$	R per sec
Modulator input	$k_1, k_2, \ldots$	$k_i = (1, \ldots, m)$	r per sec
Waveform	$s_{k_1}(t), s_{k_2}\left(t - \dfrac{1}{r}\right), \ldots$		

Each k_i is a function of K information bits

$$s_i(t) = \mathrm{Re}\,[u_i(t)\exp j2\pi f_0 t]$$

Fig. 4.2   Encoder-modulator sequences.

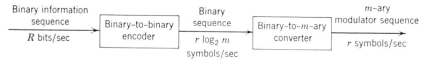

Fig. 4.3 Symbol alphabets and rates in encoder.

the input and output symbol rates of the binary-to-binary encoder are equal and the complete encoder functions only to transform the original binary information sequence into an m-ary sequence. When r exceeds the right member of (4.1), the encoder introduces redundancy into its output sequence.

Another parameter of the encoder is K, the number of information bits that enter into the determination of each m-ary encoder output. The quantity K is usually called the constraint length, in information bits, of the encoder. If no redundancy is employed, K equals $\log_2 m$, but in general both r and K are design parameters and the value of K is often best chosen to be much greater than $\log_2 m$.

The modulator accepts the m-ary sequence from the encoder as its input and transforms it into a sequence of m-ary waveforms for transmission through the channel. With the exception that we discuss in Section 4.2, the waveforms are selected from the time translates of m basic waveforms $s_i(t)$, $i = 1, \ldots, m$, which specify the modulator (see Fig. 4.2). We denote the complex envelopes of these waveforms by $u_i(t)$, $i = 1, \ldots, m$, and suppose that they possess unit norms in accord with (2.13).

Waveform Processing

The received process that results from the transmitted waveform sequence is processed by the receiver so as to determine the encoder input information sequence. The nature of the desired processing depends upon the measure of system performance, or fidelity, that is employed. When the objective is to minimize the system error probability, as it is here, the determination should be made by "deciding" that the information sequence is that one which is most probable, given the specific waveform that was received. When the information symbols are equiprobable and statistically independent, as we suppose, an equivalent decision rule is: decide that the information sequence is that one for which the (appropriately defined) conditional probability of the received process, given the transmitted sequence, is maximum.

As indicated in Fig. 4.1, we choose to divide the waveform processing into two parts that are performed by the demodulator and decoder, respectively. Broadly stated, the demodulator extracts from the received waveform all of that information relevant to a determination of the encoder input sequence, and the decoder then operates upon this information to determine the sequence. The distinction between the operations is that the demodulator processing is determined by the modulator and the channel but not by the particular encoder employed, whereas the decoder processing is determined by the encoder and by the joint statistics of the modulator input and the demodulator output.

At this point we could determine the form of the demodulator and decoder which would minimize the error probability for the system of Fig. 4.1. However, in general, it is impossible to determine the error probability of the resulting system. Therefore we confine ourselves to an important but restricted situation for which we can develop rather complete results. Roughly stated we consider those situations in which an unlimited bandwidth is available—although perhaps not used. Precisely stated, we consider those situations in which the system of Fig. 4.1 can be viewed as a diversity system of the form shown in Figs. 4.5 and 4.9. Rather than first describing the general optimum receiver and then imposing the necessary constraints piecemeal, we first present all of these constraints and then return to the receiver's structure.

4.2   WAVEFORM CONSTRAINTS

We are already constraining ourselves to systems that communicate messages rather than waveforms, that is, digital-data transmission rather than analog modulation of speech. Our other constraints fall in two categories. One category pertains to a single transmission of a basic modulator waveform and ensures a certain symmetry in the statistics of the demodulator outputs. The other category pertains to sequences of transmitted waveforms and precludes the occurrence of intersymbol interference and memory. Both kinds of constraints can be satisfied whenever adequate bandwidth is available.

Basic Modulator Waveforms

Roughly stated, we restrict ourselves to equal-energy orthogonal basic modulator waveforms that retain their orthogonality after transmission through the channel. Precisely stated, the different random processes that result from the transmission of different modulator waveforms possess orthogonal sample functions. This is equivalent to the requirement that the complex correlation functions of the channel outputs in response to different modulator waveforms have orthogonal eigenfunctions. That is, if $R_k(t, \tau)$ denotes the complex correlation function associated with the kth modulator waveform and if $\varphi_{jk}(t)$, $j = 1, \ldots,$ denotes the jth eigenfunction of $R_k(t, \tau)$ we require that

$$\int \varphi_{ik}(t)\varphi_{nq}^*(t) \, dt = 0 \tag{4.2}$$

for all i and n whenever k differs from q.

The constraint is significant when the available transmission bandwidth is limited. However the existing evidence suggests that the waveforms we consider yield the best possible performance in the absence of a bandwidth constraint [5, 6]. We impose the constraint because it renders the analytical problem tractable and because it corresponds to a condition that is often satisfied in practice. Moreover the results we obtain provide a partial introduction to the analysis of more general systems.

In the interests of simplicity, we usually also suppose that the different random processes all have the same bandwidth and that their complex correlation functions $R_k(t, \tau)$ all possess the same set of eigenvalues. These are not important constraints since our subsequent analysis is easily extended to systems that do not satisfy them. However their use leads to simpler expressions than are otherwise possible.

Frequently it is convenient to focus upon some specific set of waveforms that satisfy the constraints we are imposing. This is particularly true when we examine the implications of various mathematical expressions. We now introduce such a set: frequency position modulation. Its introduction does not constitute an additional constraint.

In frequency-position-modulation systems the complex envelopes of the m modulator waveforms are

$$u_i(t) = u(t) \exp j2\pi \, \Delta(i - 1)t, \qquad i = 1, \ldots, m \tag{4.3}$$

The quantity Δ is the frequency offset (in Hertz) between adjacent waveforms. We call $u(t)$ the *basic complex envelope* of the modulator.

It follows from (2.16) that the complex correlation function $R_k(t, \tau)$ associated with the kth of the modulator waveforms is given by the expression

$$R_k(t, \tau) = R(t, \tau) \exp j2\pi \, \Delta(k - 1)(t - \tau), \tag{4.4a}$$

where $R(t, \tau)$ is the complex correlation function of the output in response to the basic complex envelope $u(t)$, that is,

$$R(t, \tau) = \iint \sigma(r, f)u(t - r)u^*(\tau - r) \exp j2\pi f(\tau - t) \, dr \, df \tag{4.4b}$$

It is easy to show that the eigenvalues and eigenfunctions of $R_k(t, \tau)$ are, respectively, λ_i and $\varphi_i(t) \exp j2\pi \, \Delta(k - 1)t$, $i = 1, \ldots,$ where the λ_i and the

$\varphi_i(t)$ are the eigenvalues and the eigenfunctions of $R(t, \tau)$. Therefore the condition of (4.2) becomes

$$\int \varphi_i(t)\varphi_n^*(t) \exp j2\pi \, \Delta(k - q)t \, dt = 0 \qquad (4.5)$$

for all unequal values of k and q and for all values of i and n.

Equation 4.5 can be satisfied to any desired degree of precision by making Δ sufficiently large [7]. Moreover, in practice, exceedingly large values are not required—values that exceed the sum of W, the bandwidth of $u(t)$, and B, the channel frequency dispersion, should approximately suffice. The character of the approximation is evident from the following discussion.

According to (3.21), the channel output is approximately contained within a bandwidth of about $W + B$ Hz. This implies that the eigenfunctions $\varphi_i(t) \exp j2\pi\Delta(k - 1)t$ are approximately bandlimited to this same region. Of course, the location of the band depends upon both the carrier frequency and also upon the value of $k\Delta$, but if Δ exceeds $W + B$ the bands for different values of k do not overlap each other appreciably. Consequently $\varphi_i(t) \exp j2\pi\Delta(k - 1)t$ and $\varphi_n(t) \exp j2\pi\Delta(q - 1)t$ are approximately bandlimited to disjoint frequency bands if k differs from q, a condition which implies that the left member of (4.5) approximately vanishes.

Intersymbol Effects

Our second category of constraints pertains to the interference and statistical dependence that the channel may introduce into its output in response to successive transmitted waveforms. We limit our attention to situations in which there is no such interference of dependence. This situation is sometimes described by saying that intersymbol interference and memory are absent.

The constraint upon interference is quite realistic since it is satisfied by almost all fading dispersive communication systems. Unlike the previous constraints, which serve only to simplify the analysis, the intersymbol-interference constraint also simplifies the waveform processing. It is because of the processing simplification that it is often employed in practice. The supposition concerning memory is neither as realistic nor as crucial as that concerning interference. We subsequently relax it somewhat.

It is always possible to preclude the occurrence of both intersymbol effects by proper waveform selection; however the cost in bandwidth may be quite high. Thus our suppositions are tantamount to the supposition that a large bandwidth is available. We will now state the constraints precisely and describe some means by which they can be satisfied.

To define our constraints precisely, we denote the transmitted waveform sequence by $s_{k_1}(t)$, $s_{k_2}(t - 1/r)$, $s_{k_3}(t - 2/r)$, ..., as in Fig. 4.2, and represent the channel output, before the addition of noise, as

$$y(t) = \sum_i y_i(t), \tag{4.6}$$

where $y_i(t)$ is the response to $s_{k_i}(t - (i - 1)/r)$. Since the channel is linear, we are certainly free to represent the sequence in this way, but it may be impossible to recover each of the $y_i(t)$ from the output sequence. However if the $y_i(t)$ are disjoint in time, or in frequency, they are certainly recoverable. More generally, they are recoverable if the $y_i(t)$ are orthogonal to each other, that is, if the sample functions of the different responses are orthogonal. Our intersymbol-interference constraint is that they be orthogonal. Our intersymbol-memory constraint is that the $y_i(t)$ be statistically independent of each other when the transmitted waveform sequence is given.

To illustrate the nature of the constraints, we consider some specific means of satisfying them. There are other schemes that could be employed [8–11].

Let us suppose that the basic modulator complex envelope $u(t)$ has a duration of T sec and a bandwidth of W Hz. Let us further suppose that the time and frequency dispersion of the channel are L and B, respectively. As discussed in Section 3.4, the channel-output waveform possesses a duration of about $L + T$ sec and a bandwidth of about $W + B$ Hz. Since the waveforms are transmitted at a rate of r per sec, the channel responses to two successive input waveforms do not overlap appreciably, and hence there is no intersymbol interference, if r^{-1} exceeds $L + T$. Thus one way to avoid intersymbol interference is to operate at a repetition rate r that does not exceed $(T + L)^{-1}$.

We further recall that, to a first approximation, samples of the received process separated by more than about B^{-1} sec are statistically independent of each other. Consequently the channel responses to two successive transmitted waveforms are statistically independent of each other if the responses are separated in time by more than B^{-1} sec. Alternately the responses are approximately statistically independent if the transmitted waveforms are separated by at least $L + B^{-1}$ sec, that is, if

$$r < \frac{1}{T + L + B^{-1}}. \tag{4.7}$$

By the observation of the preceding paragraph, the condition of (4.7) also precludes the occurrence of intersymbol interference. Thus one technique for precluding intersymbol effects is to choose r so that it satisfies (4.7).

The acceptability of the aforestated technique depends upon the values of B and L and upon the required data rate of the system. Since there are situations in which the required rate requires the use of values of r larger than the right member of (4.7), another means of avoiding intersymbol effects is needed. One alternative is to introduce a frequency shift between successive transmissions. For example, rather than ensuring that successive channel outputs are disjoint in time, one ensures that they are disjoint in frequency. This can be accomplished by the method illustrated in Fig. 4.4.

As illustrated in the figure, the carrier frequency of the transmitter is increased by an amount Ω every r^{-1} sec until a total of $J-1$ increases are made. The carrier frequency is then returned to its original value, and the entire process is repeated. To ensure that no interference occurs between successive waveforms, the value of Ω is chosen so that the spectra of the successive responses do not overlap regardless of the modulator-input sequence. To ensure that there is no interference between successive blocks of J waveforms, J is chosen so that J/r exceeds the time duration of the response to a single waveform. With J and Ω so chosen, we can operate with any value of r without suffering intersymbol interference. For frequency position modulation (4.3), the required values of J and Ω may be estimated from the expressions

$$J \geq r(L + T) \tag{4.8a}$$

and

$$\Omega \geq m(W + B). \tag{4.8b}$$

Frequency stepping can also be employed to obtain statistical independence between the responses to successive transmissions. In particular, successive responses in a frequency-stepping system are approximately statistically independent of each other if Ω exceeds $W + B + L^{-1}$. Precisely stated, they

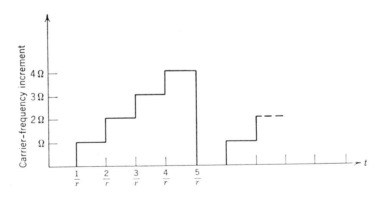

Fig. 4.4   Time-frequency diagram for frequency stepping $J = 5$.

are independent if the frequency separation between the spectra of the successive responses exceeds the coherence bandwidth of the channel. In order to ensure that all of the channel responses are statistically independent of each other, it is also necessary that the transmissions that take place on the same frequency be adequately separated in time. This can be approximately accomplished by choosing the system parameters of Fig. 4.4 so that J/r exceeds $T + L + B^{-1}$.

We have now established some means by which intersymbol interference and memory may be avoided. The supposition that there is no interference is a crucial one that underlies all of the subsequent analysis. In some instances this supposition is satisfied for all signals of interest, for example, when the channel is only frequency dispersive. There are situations, however, in which it can be satisfied only by severely restricting the value of r or by employing a frequency-stepping system with a very large value of J. That is, the elimination of intersymbol interference may impose a severe limitation upon the rate of the system unless a very large bandwidth is employed. When a high rate with limited bandwidth is required, other techniques become preferable (see Section 4.5). In contrast to the supposition of no intersymbol interference, the supposition of no memory is not a crucial one and is relaxed subsequently.

We have now introduced all of the constraints that we shall impose. We next employ them to reduce the system of Fig. 4.1 to that of Figs. 4.5 and 4.9. As a first step in that direction we determine the structure of the optimum receiver.

4.3 THE OPTIMUM RECEIVER

As stated previously, the function of the receiver is to determine the original information sequence from the received process in a manner that minimizes the system error probability. This implies that the receiver should be a maximum a posteriori probability computer or, since all information sequences are equiprobable, a likelihood-ratio computer [12–17]. That is, the receiver should compute the conditional probability of the received waveform conditioned upon the various possible information sequences and decide that the transmitted sequence is that one for which this probability is greatest.

The fraction of the received process that should be employed in the determination of an information symbol, or bit, is determined in part by the detailed definition of the error probability and in part by the character of the modulator waveforms. All of the problems we consider essentially reduce to the transmission of a single waveform, that is, to the "one-shot" communication problem. Therefore we first determine the optimum receiver for that problem.

Single Transmissions

The conditional probabilities required by the optimum receiver cannot be defined directly upon the received process. Therefore it is necessary to introduce a set of random variables, or observables, which can be extracted from the received process and which retain all of the data required by the receiver. After these observables are determined, their conditional-probability density is obtained and the form of the optimum receiver determined.

Observables: We find here that, with some modification, the variables involved in the diversity representation of Chapter 2 constitute an appropriate set of observables. To arrive at that conclusion, we first suppose that there is no additive noise and then account for that part of the noise which is relevant to our problem. For concreteness we suppose that the transmission occurs during the first time interval of Fig. 4.2.

The supposition that there is no additive noise implies that the received process may be expressed in the series form given by (2.46) provided that we know which waveform is transmitted. By virtue of the observation preceding (4.5), when the kth waveform is transmitted this series becomes

$$r(t) = y(t) = \sqrt{2E_r}\ \mathrm{Re}\left[\sum_i r_{ik}\, \varphi_i(t) \exp j\tilde{\omega}_k t\right] \qquad (4.9a)$$

where

$$r_{ik} = \left(\frac{2}{E_r}\right)^{1/2} \int r(t)\varphi_i^*(t) \exp -j\tilde{\omega}_k t\, dt, \qquad (4.9b)$$

$$\tilde{\omega}_k = 2\pi[f_o + (k-1)\Delta] \qquad (4.9c)$$

and the $\varphi_i(t)$ are the eigenfunctions of the complex correlation function $R(t, \tau)$ associated with the basic complex envelope (4.4). The quantity E_r is the average received energy per transmitted waveform.

Equations 4.9 are not suitable for our purposes because they presume a knowledge of which waveform is transmitted, whereas we require an expression that is valid in the absence of such knowledge. Although this difficulty can be remedied in a variety of ways that do not involve constraints upon the modulator waveform, the constraints we are imposing do permit a particularly simple approach.

Specifically the orthogonality condition of (4.5) implies that, no matter which waveform is transmitted,

$$r(t) = \sqrt{2E_r}\ \mathrm{Re}\left[\sum_{n=1}^{m} \sum_i r_{in}\, \varphi_i(t) \exp j\tilde{\omega}_n t\right], \qquad (4.10a)$$

where, as before,

$$r_{in} = \left(\frac{2}{E_r}\right)^{1/2} \int r(t)\varphi_i^*(t) \exp -j\tilde{\omega}_n t \, dt \qquad (4.10b)$$

and $\tilde{\omega}_n$ is defined by (4.9c). The validity of this expression follows from (4.9) upon noting that, when the kth waveform is transmitted,

$$r_{in} = 0, \qquad \text{for} \quad n \neq k, \qquad (4.11)$$

by virtue of (4.5).

It is obvious from (4.10) that the complex random variables r_{in} provide a complete representation for the received process in the absence of noise; that is, they completely determine, and are determined by, the process. But we are interested in the received process when it is corrupted by additive noise, and the expansion of (4.10a) is not then valid. However it can be shown that the variables r_{in} do preserve all of the *relevant* information contained in the received process. Precisely stated, even in the presence of white Gaussian noise, a receiver operating only upon the r_{in} defined by (4.10b) can be made to perform as well as one operating upon all of $r(t)$. This observation is an example of a result that has been called the theorem of irrelevance [14]. It is invoked implicitly, or explicitly, in almost all discussions of diversity systems and fading channels [18–26]. The only exceptions are the treatments employing either measure theory or the coordinate free methods of reproducing kernels in Hilbert space [27–31]. These two approaches are more rigorous than that which we present here but they lead to comparable results.

The theorem of irrelevance is of interest to us because it implies that the optimum receiver need evaluate only the joint conditional probabilities of the observables r_{in}, $n = 1, \ldots, m$, $i = 1, \ldots$. We turn now to an evaluation of those probabilities.

Conditional Probabilities: Let us suppose that the kth waveform is transmitted and also, initially, suppose that there is no additive noise. Then, by virtue of (4.11), r_{in} vanishes for n not equal to k and, by virtue of the discussion involving (2.48), the r_{ik} are statistically independent zero-mean complex Gaussian random variables that are uncorrelated with their complex conjugates; that is,

$$\overline{r_{ik}} = 0, \qquad \text{for all } i, \qquad (4.12a)$$

$$\overline{r_{ik} r_{jk}} = 0, \qquad \text{for all } i \text{ and } j, \qquad (4.12b)$$

$$\overline{r_{ik} r_{jk}^*} = 0, \qquad \text{for } i \neq j. \qquad (4.12c)$$

Moreover the variance $\overline{|r_{ik}|^2}$ of r_{ik} equals the ith eigenvalue of the complex correlation function of the received process that results from the transmission of the kth waveform. For the frequency position modulation that we are considering, this eigenvalue equals the ith eigenvalue λ_i of the basic complex correlation function $R(t, \tau)$ of (4.4b); that is,

$$\overline{|r_{ik}|^2} = \lambda_i \qquad (4.12d)$$

when the kth waveform is transmitted.

It follows from (4.10b) that the effect of adding a white Gaussian noise $n(t)$ to the received process is to add a random variable

$$n_{in} = \left(\frac{2}{E_r}\right)^{1/2} \int n(t)\varphi_i^*(t) \exp -j\tilde{\omega}_n t \, dt \qquad (4.13)$$

to the values of the r_{in} that result in the absence of noise. Equation 4.13 and the properties of the $\varphi_i^*(t)$ imply that the n_{in} are statistically independent zero-mean Gaussian random variables uncorrelated with their complex conjugates and possessing the common variance N_0/E_r; that is,

$$\overline{n_{in}} = 0, \qquad \text{all } i \text{ and } n, \qquad (4.14a)$$

$$\overline{n_{in}n_{pq}} = 0, \qquad \text{all } i, n, p, \text{ and } q, \qquad (4.14b)$$

$$\overline{n_{in}n_{pq}^*} = 0, \qquad \text{unless } i = p \text{ and } n = q, \qquad (4.14c)$$

$$\overline{|n_{in}|^2} = \frac{1}{\alpha}, \qquad (4.14d)$$

where we have introduced

$$\alpha = \frac{E_r}{N_0}, \qquad (4.14e)$$

which is the average-received-energy–to–noise-power-density ratio per transmitted modulator waveform.†

Equations 4.12 through 4.14, with the comments that precede them, lead us to conclude that, in the presence of noise, the r_{in} are jointly Gaussian, zero-mean, independent random variables when the transmitted waveform is known. In particular, when the kth waveform is transmitted,

$$\overline{r_{in}} = 0, \qquad \text{all } i \text{ and } n, \qquad (4.15a)$$

$$\overline{r_{in}r_{pq}} = 0, \qquad \text{all } i, n, p, \text{ and } q, \qquad (4.15b)$$

$$\overline{r_{in}r_{pq}^*} = 0, \qquad \text{unless } i = p \text{ and } n = q, \qquad (4.15c)$$

$$\overline{|r_{in}|^2} = \alpha^{-1}, \qquad \text{if } n \neq k, \qquad (4.15d)$$

$$\overline{|r_{ik}|^2} = \lambda_i + \alpha^{-1}. \qquad (4.15e)$$

† Following common usage, we will often refer to α as the energy-to-noise ratio; it is, in fact, the ratio of total average received energy to noise power per Hz.

We note in passing that (4.15) are predicated upon the supposition that the complex correlation functions of the channel output in response to the different modulator waveforms all possess the same set of eigenvalues—as they do for frequency position modulation. To extend (4.15) to waveform sets for which the eigenvalue sets differ, it is only necessary to replace λ_i in (4.15e) by $\lambda_i{}^k$, the ith eigenvalue of the complex correlation function of the output process that results from the transmission of the kth modulator waveform.

The conditional probability density $p(\mathbf{r}\,|\,k)$ of the set of observables r_{in}, given that the kth waveform is transmitted, follows easily from (4.15) and the associated discussion. Specifically,

$$p(\mathbf{r}\,|\,k) = \left\{ \prod_i \left[\left(\frac{\alpha}{2\pi}\right)^m \exp -\frac{\alpha}{2} \sum_n |r_{in}|^2 \right] \right\}$$

$$\times \left\{ \prod_i \left[(1 + \alpha\lambda_i)^{-1} \exp \frac{\alpha}{2} \frac{\alpha\lambda_i}{1 + \alpha\lambda_i} |r_{ik}|^2 \right] \right\}, \qquad (4.16)$$

where the r_{in} are defined by (4.10b), α is the energy-to-noise ratio defined by (4.14e) and the λ_i are the eigenvalues of the complex correlation function associated with $u(t)$.

It is clear from (4.16) that the value of k that maximizes $p(\mathbf{r}\,|\,k)$ also maximizes the quantity f_k defined as

$$f_k = \alpha \sum_i \frac{\alpha\lambda_i}{1 + \alpha\lambda_i} |r_{ik}|^2, \qquad (4.17a)$$

where, as before,

$$r_{ik} = \left(\frac{2}{E_r}\right)^{1/2} \int r(t)\varphi_i^*(t) \exp -j\tilde{\omega}_k t\, dt \qquad (4.17b)$$

with

$$\alpha = \frac{E_r}{N_0} \qquad (4.17c)$$

and

$$\tilde{\omega}_k = 2\pi[f_o + (k-1)\Delta]. \qquad (4.17d)$$

Consequently the optimum receiver need evaluate only the quantities f_k and decide that the transmitted waveform is that one corresponding to the largest f_k. That is, the transmitted waveform is taken to be that one, say j, for which

$$f_j \geq f_i, \qquad i = 1, \ldots, m. \qquad (4.18)$$

This decision rule yields ambiguous results whenever (4.18) is satisfied for more than one value of j, but the error probability of the system is not affected by the way in which these ambiguities are resolved.

Diversity Interpretation: It is instructive to describe the single-transmission system directly in terms of the observables r_{in}. These observables are completely described by (4.15) and may be envisaged as the outputs of the system shown in Fig. 4.5. In the figure, s_k is equal to unity when the kth modulator waveform is transmitted and is equal to zero otherwise.

Figure 4.5 is the extension of the diversity representation of Chapter 2 (Fig. 2.4) to communication systems. It is also a familiar model for digital diversity communication systems employing orthogonal waveforms with an energy-to-noise ratio of $\alpha\lambda_i$ for the ith diversity path [32–44]. Since the

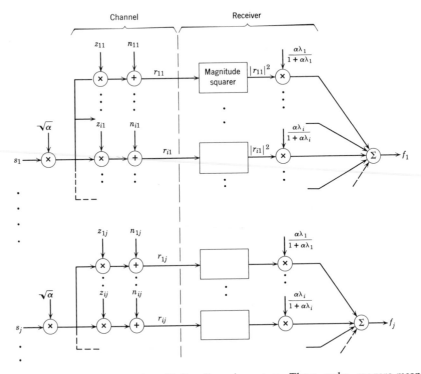

Fig. 4.5 Diversity interpretation of fading dispersive system. The z_{ij} and n_{ij} are zero-mean complex Gaussian random variables. All variables are statistically independent of each other and $\overline{z_{ij}}^2 = \overline{n_{ij}}^2 = 0$, $\overline{|z_{ij}|^2} = \lambda_i$, $\overline{|n_{ij}|^2} = \alpha^{-1}$.

optimum receiver for our channel involves only the r_{in}, the system we are considering performs as the diversity system of the figure. This identification proves to be a fruitful one. It also prompts us to refer to the $\alpha\lambda_i$ as the energy-to-noise ratios of the diversity paths and to the λ_i as fractional path strengths (recall that $\sum_i \lambda_i = 1$).

Thus far we have assumed that a single one of the modulator waveforms is transmitted. However our results are easily extended to the transmission of sequences when there is no intersymbol memory or interference. We turn now to that extension.

Sequences

As in (4.6), we represent the channel output, before the addition of noise, in the form

$$y(t) = \sum_w y_w(t), \tag{4.19}$$

where $y_w(t)$ is the output in response to the wth waveform in the transmitted sequence. We then invoke the constraint of no intersymbol interference, which implies that each of the $y_w(t)$ can be recovered from $y(t)$.

The constraint is significant because it enables us to form the observables for $y(t)$ from those associated with the separate $y_w(t)$. To this end we appeal to (4.10) to conclude that the $y_w(t)$, and hence $y(t)$, can be constructed from the observables

$$r_{inw} = \left(\frac{2}{E_r}\right)^{1/2} \int y_w(t)\varphi_i^*\left(t - \frac{w}{r}\right) \exp\, -j\tilde{\omega}_n\left(t - w/r\right)\, dt,$$

$$i = 1, \ldots, n = 1, \ldots, m, w = 0, 1, \ldots. \tag{4.20}$$

In this expression the $\varphi_i(t - w/r)\exp j\tilde{\omega}_n(t - w/r)$, $i = 1, \ldots$, are, in fact, the eigenfunctions of the complex correlation function associated with $y_w(t)$ when the wth waveform in the transmitted sequence is $s_n(t - w/r)$. Note that the r_{inw} for $w = 0$ are just the r_{in} of (4.10).

We next note that the r_{inw} can be extracted directly from $y(t)$ without first resolving it into the separate responses $y_w(t)$. Specifically they can be evaluated from (4.20) with $y_w(t)$ replaced by $y(t)$. This replacement is possible because the noninterference constraint ensures that the sample functions of the different $y_w(t)$ are orthogonal to each other and this orthogonality, in conjunction with (4.10), implies that the $\varphi_i^*(t - w/r)\exp\, -j\tilde{\omega}_n(t - w/r)$ are orthogonal to the sample functions of $y_v(t)$ for v not equal to w. Consequently if we momentarily suppose that no noise is present so that $r(t)$ equals $y(t)$, we have

$$r_{inw} = \left(\frac{2}{E_r}\right)^{1/2} \int r(t)\varphi_i^*\left(t - \frac{w}{r}\right) \exp -j\tilde{\omega}_n (t - w/r) \, dt,$$

$$i = 1, \ldots, n = 1, \ldots, m, \, w = 0, 1, \ldots. \quad (4.21)$$

In summary, $y(t)$ can be reconstructed from the variables r_{inw} of (4.21), and the evaluation of these variables does not require knowledge of which waveform is transmitted. Moreover the r_{inw} contain all of the information relevant to a determination of the transmitted waveform sequence even when additive white Gaussian noise is present. Thus they are the only quantities that the optimum receiver need extract from the received waveform $r(t)$.

It remains to determine the conditional probabilities of the r_{inw} for each possible information sequence. In general these probabilities may be complicated functions of the statistical dependencies that the channel can introduce. However for the memoryless situation that we are considering they become especially simple because the r_{inw} are all statistically independent of each other given the information sequence. This independence can be established by an argument similar to that involving (4.12) through (4.15), in conjunction with the observation that all of the $\varphi_i(t - w/r) \exp j\tilde{\omega}_n(t - w/r)$ are orthogonal to each other.

The independence of the r_{inw} implies that

$$p(\mathbf{r} \mid S(t)) = \prod_w p(\mathbf{r}_w \mid k_w), \quad (4.22)$$

where $p(\mathbf{r} \mid S(t))$ is the conditional probability density of the entire set of r_{inw}, given the transmitted waveform sequence $S(t)$, and $p(\mathbf{r}_w \mid k_w)$ is the conditional density of the set of r_{inw} for a fixed w, given the index k_w of the wth waveform in the transmitted sequence.

Finally we note that the probability densities appearing in the right member of (4.22) are given by (4.16) with $\mathbf{r}$ and k replaced by $\mathbf{r}_w$ and k_w. Consequently the left member of (4.22) can be expressed in the form

$$p(\mathbf{r} \mid S(t)) = F(\mathbf{r}) \exp \frac{1}{2} \sum_w f_{k_w w}, \quad (4.23a)$$

where $F(\mathbf{r})$ is independent of the transmitted waveform sequence and

$$f_{nw} = \alpha \sum_i \frac{\alpha \lambda_i}{1 + \alpha \lambda_i} |r_{inw}|^2. \quad (4.23b)$$

Clearly the maximization of the left member of (4.23a) over all the sequences that are generated by the transmitter is equivalent to the maximization of

$$\sum_w f_{k_w w} \quad (4.24)$$

over the set of m-ary sequences $k_0, k_1, \ldots$, that are generated by the encoder, that is, over the set of code words.

The Demodulator

At this juncture the distinction between demodulation and decoding emerges naturally. The determination of the f_{nw} does not require any knowledge of the encoder procedure, whereas, given the f_{nw}, $n = 1, \ldots, m$, $w = 0$, $1, \ldots$, the maximization of (4.24) requires only a knowledge of that procedure. Thus we define the demodulator as that portion of the receiver which computes the f_{nw} and define the decoder as that portion which performs the maximization of (4.24). Since the receiver we are considering is optimum in the sense of error probability, we frequently use the terms optimum demodulator and optimum decoder.

The structure of the optimum demodulator is influenced by the techniques that are employed to preclude intersymbol effects. If no special techniques are employed, the evaluation of the f_{nw} can be implemented as shown in Fig. 4.6. As indicated in the figure, a set of m parallel, or branch, demodulators is employed in the evaluation of f_{nw}, $n = 1, \ldots, m$. Each branch demodulator consists of a basic demodulator element preceded by a multiplier that, in essence, shifts the carrier frequency of the received process to the passband of the demodulator. This basic demodulator can be implemented as either a set of correlation devices (Fig. 4.7a) or a set of matched filters (Fig. 4.7b). Of course, in practice the receiver employs only a single set of m branch demodulators which are reset for the evaluation of successive f_{nw}. This is

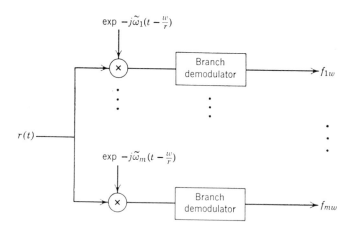

Fig. 4.6 Evaluation of the f_{nw}, $n = 1, \ldots, m$.

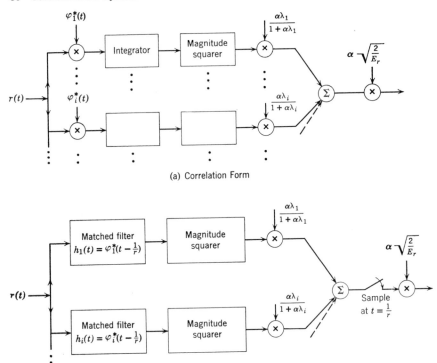

(a) Correlation Form

(b) Matched Filter Form

Fig. 4.7   Branch demodulator implementation for $w = 0$.

easily accomplished with the matched filter realization of the basic demodulator.

Some modification of the receiver structure is necessary when frequency stepping is used to eliminate interference and memory. Since the channel responses to J successive transmissions may influence the received waveform at any given time and since each of the responses is to be processed separately, J sets of m branch demodulators are required. Each set is preceded by a "multiplier" that serves to undo the frequency stepping introduced at the transmitter. One possible implementation is shown in Fig. 4.8.

It is evident from Fig. 4.6 that the principal element of the demodulator is the branch demodulator. We find in Chapter 7 that the basic demodulator can sometimes be implemented in relatively simple forms. Our present objective was merely to demonstrate that the operations performed by the optimum demodulator reduce, in essence, to those performed by the basic branch demodulator. That we have now done. We next determine the statistics of the demodulator output.

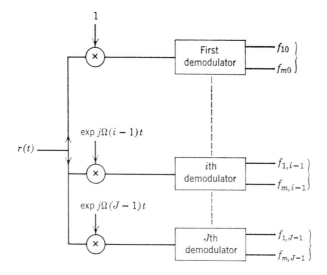

Fig. 4.8 Demodulator for frequency stepping. The ith demodulator computes f_{nw}, $n = 1, \ldots, m$, $w = i - 1$, $J + i - 1$, $2J + i - 1, \ldots$, and delivers the results as outputs at times i/r, $(J + i)/r$, $(2J + i)/r, \ldots$.

We seek the joint conditional statistics of the f_{nw}, $n = 1, \ldots, m$, $w = 0$, $1, \ldots$, given the information sequence or equivalently the transmitted waveform sequence. In the absence of intersymbol memory the r_{inw} that enter into the determination of the different f_{nw} are independent [recall (4.22)] and hence so also are the different f_{nw}. Moreover since there is no intersymbol interference, the conditional density of f_{nw} depends only upon the wth waveform in the transmitted sequence. Consequently it suffices to determine the statistics of each f_{nw} conditioned upon the wth transmission. That is, we need only consider the single transmission problem. These statistics possess strong symmetry properties as a result of the waveform constraint of Section 4.2.

First, given the knowledge of the transmitted waveform, the outputs of the different branch demodulators are statistically independent of each other. Second, the probability density of the "correct" branch demodulator does not depend upon which branch is correct. By the "correct" branch demodulator we mean that one whose index agrees with the transmitted waveform; for example, when the kth waveform is transmitted, the kth branch is "correct" and all others are "incorrect." Finally, the $m - 1$ incorrect branches all possess the same probability density.

Roughly stated, these properties follow from the facts that (1) the "signal" contributes to the output of only the correct demodulator, (2) all of the "signals" are "similar," and (3) the demodulator passbands are disjoint. Precisely stated, the properties are consequences of (4.16) and (4.22) and their associated discussion. In particular, given the knowledge of which waveform is transmitted, all of the variables r_{inw} are independent of each other and hence so also are the demodulator outputs. Moreover the statistics of the variables that contribute to the output of the correct branch do not depend upon which branch is correct; hence neither does the probability density of the output. Finally, the variables that contribute to the incorrect branch outputs are all identically distributed and so also are the outputs.

The probability densities of the "correct" and "incorrect" branch demodulator outputs cannot be expressed in any convenient form, but their moment-generating functions can be. These functions, which are in essence the Laplace transforms of the probability densities, are required in the sequel, and so we introduce them here.

We denote the moment-generating functions of the "correct" and "incorrect" branch demodulator by $g_0(s)$ and $g_1(s)$, respectively. They are defined by the expressions

$$g_0(s) = \overline{\exp sf_k} \qquad (4.25a)$$

and

$$g_1(s) = \overline{\exp sf_j}, \qquad (4.25b)$$

where f_k is the output of the correct branch and f_j is the output of any one of the incorrect branches. These functions, which exist for all values of s that do not exceed zero, can be shown to be given by the following expressions [45–47]:

$$g_0(s) = \prod_i (1 - s\alpha\lambda_i)^{-1}, \qquad (4.26a)$$

$$g_1(s) = \prod_i \left(1 - s\,\frac{\alpha\lambda_i}{1 + \alpha\lambda_i}\right)^{-1}. \qquad (4.26b)$$

As before, the λ_i are the eigenvalues of $R(t, \tau)$, the complex correlation function associated with the basic complex envelope $u(t)$, and α is the average-received-energy–to–noise ratio.

It is frequently convenient to present the demodulator output statistics in the form of Fig. 4.9. There we replace the original modulator-channel-demodulator by a new "channel." The input to this new channel is the m-ary input sequence to the modulator and its output is the (vector) output sequence of the demodulator. The new channel is memoryless in that each vector output $\mathbf{f}_w = (f_{1w}, \ldots, f_{mw})$ is statistically dependent only upon the corresponding channel input. The statistical description of the new channel may be given either as in the figure or as in the diversity interpretation of Fig. 4.5.

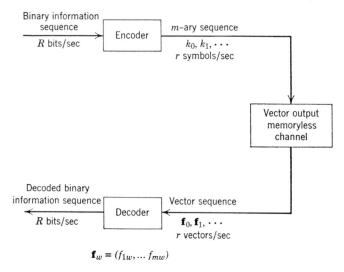

$$\mathbf{f}_w = (f_{1w}, \dots f_{mw})$$

Fig. 4.9 Reduced description of fading dispersive communication system. Given the channel input, say k_w, the elements of $\mathbf{f}_w$ are statistically independent random variables. The moment-generating function of $fk_w w$ is $g_o(s) = \prod_i (1 - \alpha s \lambda_i)^{-1}$, and the moment-generating function of every other element is $g_1(s) = \prod_i [1 - \alpha s \lambda_i/(1 + \alpha \lambda_i)]^{-1}$. The average energy-to-noise ratio per channel use is α, and the fractional path strengths are λ_i.

4.4 SYSTEM PERFORMANCE

Thus far we have been primarily concerned with the form of the communication system; we now consider its performance. Our objectives are to introduce appropriate definitions of the system error probability, to obtain expressions from which this probability may be evaluated, and to reexamine the need for the restriction that there be no intersymbol memory.

The performance of the system of Figs. 4.1 and 4.9 can be described by several different error probabilities. For example, there is the "bit" error probability, the baud error probability, and the "code word" error probability. The choice between them is determined in part by the coding employed and in part by expediency.

Uncoded Systems

In the absence of coding, every conceivable *m*-ary sequence represents a possible information sequence. Thus the maximization of (4.24) reduces to the maximization of f_{nw} over n for each w and the decoder merely decides that the *w*th successive (*m*-ary) input to the modulator is that one (say k) for which

$$f_{kw} \geq f_{nw}, \qquad n = 1, \dots, m. \tag{4.27}$$

As before, the possible ambiguities occur with zero probability and are of no consequence.

It is most appropriate to describe the performance of uncoded systems by the (baud) error probability $P(\varepsilon)$, that any particular m-ary decision is incorrect. Since the channel is memoryless, errors in successive bauds are independent of each other, and the probability of any sequence of baud errors can be easily determined from $P(\varepsilon)$. Moreover it turns out that all possible errors are equiprobable so that the bit error probability is $\frac{1}{2}[m/(m-1)]P(\varepsilon)$.

To obtain an expression for $P(\varepsilon)$, we note that, since all modulator inputs are equiprobable,

$$P(\varepsilon) = \frac{1}{m} \sum_{k=1}^{m} P(\varepsilon \mid k), \tag{4.28a}$$

where $P(\varepsilon \mid k)$ is the conditional probability of a baud error when the modulator input is k. Clearly $P(\varepsilon \mid k)$ is just the probability that one or more of the f_{nw} is as large as f_{kw}, where w refers to the baud being considered; that is,

$$P(\varepsilon \mid k) = 1 - P(f_{kw} > f_{iw} \text{ all } i \neq k). \tag{4.28b}$$

We next recall that, given k, the variables $f_{1w}, \ldots, f_{mw}$ are statistically independent of each other and that the moment-generating function of f_{kw} is given by (4.26a), whereas the moment-generating function of the remaining variates is given by (4.26b). Since these statistics do not depend on k, all of the $P(\varepsilon \mid k)$ are equal and (4.28a) may be restated as

$$P(\varepsilon) = 1 - P(x > y_j, j = 1, \ldots, m-1), \tag{4.29}$$

where the variables x and $y_j, j = 1, \ldots, m-1$, are statistically independent random variables. The moment-generating function of x is $g_o(s)$, as given by (4.26a), and the moment-generating function of each y_j is $g_1(s)$, as given by (4.26b).

An exact evaluation of (4.29) is difficult if not impossible. Therefore we choose to overestimate and underestimate the right member of that equation with relatively sharp and simple bounds that are derived in Appendix 2. Although we defer evaluation of the bounds to Chapter 5, it is convenient to present them here. They are as follows:

$$P(\varepsilon) \leq P(x \leq h) + mP(h < x \leq y). \tag{4.30a}$$

Also

$$P(\varepsilon) \geq \frac{m}{4} P(x \leq h)P(y \geq h) \tag{4.30b}$$

provided that $mP(y \geq h) \leq 1$. The quantity h appearing in these expressions is a parameter that can be chosen to obtain the sharpest result; x and y are statistically independent random variables with the moment-generating functions $g_o(s)$ and $g_1(s)$, given, respectively, by (4.26a) and (4.26b).

For future reference we note that the bounds of (4.30) are valid for any demodulator that satisfies the conditions summarized in Fig. 4.9 when it is used with the decision rule of (4.27). It is only necessary to redefine the variables x and y in terms of the demodulator employed. It is also worth noting that the decision rule defined by (4.27) yields an error probability that satisfies (4.29) regardless of the a priori probabilities of the m different waveforms

One other point deserves further comment. Although we suppose here that there is no intersymbol memory, the results of this section can also be applied to channels in which memory is present. If the decision rule of (4.27) is employed with a channel that possesses memory, the baud error probability still satisfies (4.29) and (4.30). Of course successive errors no longer are statistically independent. Moreover the resulting system does not yield as small an error probability as does one that exploits the memory. However, in the interests of simplicity, such suboptimum systems are often employed.

Block-Coded Systems

When block coding is employed, each successive " block " of K information symbols is associated with a successive block of N modulator waveforms. The maximization of (4.24) then reduces to an independent maximization over each successive block, or code word. The performance of such a system is conveniently described by the probability that any particular block of output information symbols differs from the corresponding input block. This probability is called the word error probability. We denote it by $P(\varepsilon)$ and rely upon the context to distinguish between word and baud errors.

In general the evaluation of $P(\varepsilon)$ for specific "good" codes is impractical. However it is possible to prove that codes exist for which the error probability is less than a value denoted by $\overline{P(\varepsilon)}$. In certain situations it can be shown that the error probability cannot be too much less than $\overline{P(\varepsilon)}$. These results are obtained by the random coding and sphere packing arguments of coding theory. We do not pursue these matters in detail because they are treated extensively elsewhere [48–52]. We do however present the random coding result for the channel of Fig. 4.9.

    Random-coding Bound: The random-coding theorem implies that, for any value of N, a block code exists whose error probability is no larger than

$$\overline{P(\varepsilon)} \leq 2^{-(N/r)E_c}, \tag{4.31a}$$

where

$$E_c = \max_{0 \le \rho \le 1} [rE_o(\rho) - \rho R], \tag{4.31b}$$

$$E_o(\rho) = -\log_2 \left\{ \int \left[\sum_i P_i \, p(\mathbf{f}|i)^{1/(1+\rho)} \right]^{1+\rho} d\mathbf{f} \right\}. \tag{4.31c}$$

In these expressions R is the information rate in bits per second, r is the number of channel symbols per second, and $p(\mathbf{f}|k)$ is the conditional probability density of the channel-output vector given that the input symbol is k. The P_i, which are a set of probabilities over the set of m input symbols, satisfy the expression

$$\int [p(\mathbf{f}|k)]^{1/(1+\rho)} \left[\sum_i P_i \, p(\mathbf{f}|i)^{1/(1+\rho)} \right]^{\rho} d\mathbf{f}$$

$$\ge \int \left[\sum_i P_i \, p(\mathbf{f}|i)^{1/(1+\rho)} \right]^{1+\rho} d\mathbf{f} \tag{4.31d}$$

for all k with equality for those k such that $P_k \ne 0$.

Equation 4.31c can be reduced to a more convenient form by recalling that (1) the elements $f_1, \ldots, f_m$ of the channel-output vector $\mathbf{f}$ are statistically independent conditioned upon the transmission of k, (2) with the exception of f_k, all of the elements of $\mathbf{f}$ are identically distributed with the moment-generating function $g_1(s)$ of (4.26b), and (3) f_k possesses the moment-generating function $g_o(s)$ of (4.26a). Consequently

$$p(\mathbf{f}|k) = p_o(f_k) \prod_{i \ne k} p_1(f_i), \tag{4.32}$$

where $p_o(\cdot)$ is the density associated with $g_o(s)$ and $p_1(\cdot)$ is the density associated with $g_1(s)$. We also note, from (4.26), that

$$g_o(s) = \frac{g_1(s+1)}{g_1(1)}. \tag{4.33}$$

Consequently by the properties of Laplace transforms

$$p_o(f_k) = \frac{p_1(f_k) \exp f_k}{g_1(1)}. \tag{4.34}$$

When (4.32) and (4.34) are introduced into (4.31d), it is found that the resulting condition is satisfied when the P_i are taken to be equal. Substitution of these values into (4.31c) yields

$$E_o(\rho) = -\log_2 \left\{ \int \left[\prod_{i=1}^{m} p_1(f_i) \right] \left(\sum_{k=1}^{m} \exp \frac{f_k}{1+\rho} \right)^{1+\rho} \frac{1}{g_1(1)m^{1+\rho}} \, d\mathbf{f} \right\}. \tag{4.35}$$

In general, numerical techniques must be employed to evaluate (4.35). However when ρ equals unity, a simpler analytical expression can be obtained by expanding the squared summation and evaluating the integrals in terms of $g_1(\cdot)$. The result is

$$E_o(1) = -\log_2 \left[\frac{1}{m} + \left(1 - \frac{1}{m}\right) \frac{g_1^{\,2}(\tfrac{1}{2})}{g_1(1)} \right]. \tag{4.36}$$

We further discuss the performance of general block-coded systems in Chapter 5. Our objective here was simply to reduce the problem to the evaluation of (4.31) and (4.36).

It is worth noting the role of the memoryless constraint in our discussion of coding. The random coding bound of (4.31) is predicated upon the absence of such memory between the channel outputs that enter into the determination of a code word. However it may be applied to systems in which memory exists between code words, just as the baud error probability of uncoded systems can be applied to systems with memory between bauds. This situation arises when we employ scrambling, or interlacing, to preclude memory between the outputs that are associated with each code word [53]. That is, rather than coding over successive blocks of N modulator waveforms, each successive code word employs waveforms that are widely separated in the transmitted sequence. This is not an optimum way of dealing with the memory that exists, but it does permit us to utilize memoryless decoding techniques.

In addition to the general block codes to which (4.31) apply, we have occasion to consider the special class of orthogonal codes. Or perhaps in more familiar language, we consider diversity transmission schemes. Such schemes provide a valuable degree of freedom in the design of fading dispersive communication systems. They enter into our discussion of waveform design in Chapter 6.

Explicit Diversity: An orthogonal code is one in which the transmitted waveform sequences representing different code words are orthogonal to each other. For the orthogonal modulator waveforms that we are considering, a particularly simple orthogonal code of length N is obtained by repeating the transmission of a modulator waveform N times. This yields a system that is identical to an N-fold time-diversity transmission system. To distinguish the externally obtained diversity from that inherent in the basic modulator waveform, we frequently refer to the former as explicit diversity and the latter as implicit diversity.

Although useful, the distinction between the forms of diversity is somewhat vague because it depends upon the definition of the "modulator" waveforms. For example, an N-fold time-diversity system employing a basic modulator complex envelope $u(t)$ can also be viewed as a system, without explicit diversity, in which the basic modulator complex envelope is $(1/\sqrt{N})$ $\times \sum_{i=1}^{N} u(t - (i-1)/r)$ and the average received energy-to-noise ratio is $N\alpha$. To eliminate some of the ambiguity we frequently specify the basic complex envelope. Thus the system we are considering here is an N-fold explicit diversity system based upon $u(t)$.

It is apparent that the information rate of a time-diversity system decreases inversely with N for any fixed modulator. This decrease can be avoided by the use of frequency diversity. That is, rather than beginning with the basic complex envelope $u(t)$ and using time translates of $u(t)$ to obtain an effective envelope of $(1/\sqrt{N}) \sum_{i=1}^{N} u(t - (i-1)/r)$, we may use frequency translates of $u(t)$ to obtain an effective envelope $(1/\sqrt{N}) \sum_{i=1}^{N} u(t) \exp j\Omega(i-1)t$. More generally, any combination of time and frequency translates of $u(t)$ may be employed, that is, any combination of time and frequency diversity. The system error probability is not affected by the particular combination employed provided that the N translates are separated enough to ensure that their corresponding channel outputs are statistically independent and do not interfere.

It is most convenient to define and determine the error probability of an explicit diversity system in terms of the equivalent uncoded system employing the effective envelope of the preceding paragraph. The definition and results that lead to Fig. 4.9 can be applied directly to this equivalent system to obtain the error probability bounds of (4.30). Of course, the moment-generating functions of the random variables involved in that bound are determined by the eigenvalues of the complex correlation function associated with the effective envelope $(1/\sqrt{N}) \sum_{i=1}^{N} u(t - (i-1)/r)$ rather than by the eigenvalues of the complex correlation function associated with $u(t)$ itself. However it is not too difficult to express the latter in terms of the former. Specifically, if $\lambda_i, i = 1, \ldots,$ are the eigenvalues associated with $u(t)$, the eigenvalues associated with the effective envelope are

$$\underbrace{\frac{\lambda_1}{N}, \ldots, \frac{\lambda_1}{N}}_{N \text{ times}}; \underbrace{\frac{\lambda_2}{N}, \ldots, \frac{\lambda_2}{N}}_{N \text{ times}}; \ldots; \ldots.$$

The validity of this conclusion hinges upon the intersymbol constraints.

The expressions for the required moment-generating functions follow easily from the preceding comments in conjunction with (4.26). Specifically the error probability of the system satisfies (4.29) and (4.30) with $g_o(s)$ and $g_1(s)$ replaced by $g_o^N(s)$ and $g_1^N(s)$, where $g_o(s)$ and $g_1(s)$ are still given by (4.26). The λ_i in (4.26) are still the eigenvalues associated with $u(t)$, and α is the energy-to-noise ratio associated with the transmission of a single modulator waveform.

The replacement of $g_0(s)$ and $g_1(s)$ by $g_0{}^N(s)$ and $g_1{}^N(s)$ is equivalent to the addition of more diversity paths in the diversity picture of Fig. 4.5. In effect, the use of N-fold explicit diversity causes each diversity path of that figure to be reproduced N times. Thus we may think of the total diversity in the system as the product of the explicit and implicit diversity. We return to this point in Chapter 6.

Finally we note that the required independence between the constituent parts of an explicit diversity system can be obtained by scrambling, just as it can with general coded systems [53]. When scrambling is employed, successively transmitted modulator waveforms need be separated in time and frequency only enough to preclude interference between them.

Bandwidth and Complexity

Two factors cause large orthogonal alphabets to be impractical: the excessive transmission bandwidth and the receiver complexity required for their use. In the absence of coding ($v = K$), both of these quantities increase exponentially with the values of K. In fact, the receiver consists of $m = 2^K$ branch demodulators. Also, each of the transmitter waveforms possesses a bandwidth of at least about r or equivalently R/K Hz, where r^{-1} is the time allocated to the transmission of K information bits and R is the information rate in bits per second. Since there are 2^K waveforms and since the waveforms are all orthogonal, the total transmission bandwidth is about $(R/K)\,2^K$ Hz. Thus if K exceeds 10, the transmission bandwidth must be at least 100 times the information rate, and the receiver must employ more than 1,000 branch demodulators.

For the coded systems that we consider, the transmission bandwidth is roughly equal to mr, where m is the modulator alphabet size and r^{-1} is the time allotted to the transmission of a single modulator waveform. Thus K may be made arbitrarily large, for a fixed information rate R, without increasing the bandwidth requirement. Hence, it is possible to employ a large number of coded waveforms without requiring an excessive bandwidth. Although we sacrifice the orthogonality among these waveforms, we find that the sacrifice need not be costly. In fact, in Chapter 5 we find that a system employing about 16 waveforms, in conjunction with digital coding over K successive information bits, can be made to perform almost as though it employed an orthogonal alphabet of 2^K waveforms.

The use of block coding does not eliminate the exponential dependence of receiver complexity on K, but it does alter its character. Briefly stated, it shifts the complexity from the demodulator to the decoder. This shift is often desirable because it places the burden upon digital rather than analog equipment; however even the digital-processing requirements are impractical for moderately large values of K. This problem persists because the optimum receiver must evaluate, or store, at least 2^K quantities to decode each successive block of K information bits. Consequently, optimum receivers employing values of K exceeding, say, 30 are totally beyond reach for the foreseeable future.

Fortunately a number of practical suboptimum procedures have been developed that retain much of the power of a block code employed with an optimum receiver. Since these procedures are described elsewhere, we do not pursue them here [54–61]. Suffice it to say that the results we obtain for block codes are indicative of the performance that can be achieved in practical systems. This applies not only to the use of algebraic or sequential procedures to alleviate the computational problem but also to the quantization of the demodulator outputs to alleviate the data-storage problem.

One final comment is in order. Although the coding we are considering does provide one means of reducing the bandwidth requirement of large alphabet systems, it is not the most efficient means of accomplishing this reduction. Thus if bandwidth conservation is a primary consideration in a system, waveforms other than those that we are considering should be employed. Although the precise character of the "best" waveforms is not known, some insight into their characteristics is provided by the generalizations discussed in Section 4.5.

The formulation of the communication problem we wish to consider is now complete, and we can proceed with its solution. First, however, we pause to relate the restricted problem we are considering to more general problems and to other approaches of engineering interest.

4.5 GENERALIZATIONS AND EXTENSIONS

The systems we consider above are constrained in two directions: the "orthogonality" of the modulator waveforms and the absence of intersymbol effects. The orthogonality constraint is an analytic convenience that permits us to make some general statements about the performance of a class of fading dispersive systems. The intersymbol constraint, on the other hand, is less significant in the analysis of systems than it is in their implementation. That is, if the single transmission problem can be solved without the orthogonality constraint, the results can be applied to the transmission of sequences by the simple artifice of redefining the "single-transmission" interval. However the optimum receiver associated with this redefinition is exceedingly complex if intersymbol effects are present [18–24]. Thus we frequently either choose the system waveforms so as to preclude the occurrence of intersymbol effects or employ a suboptimum receiver structure. In the following discussion we briefly describe the problems that arise when these constraints are relaxed and the direction in which the development of sub-optimum receivers has proceeded.

General Single-Transmission Problem

The general single-transmission problem can always be reduced to a system involving random variables, much as Fig. 4.1 was reduced to Fig. 4.5. This is accomplished by expanding the transmitted and received waveforms in an orthonormal series with an arbitrarily chosen set of complete ortho-normal functions. If we denote the members of this set by $\chi_i(t)$, $i = 1, \ldots,$ the expansions are

$$s(t) = \sum_i s_i \chi_i(t), \qquad (4.37a)$$

$$y(t) = \sum_i y_i \chi_i(t), \qquad (4.37b)$$

and

$$r(t) = y(t) + n(t) = \sum_i r_i \chi_i(t), \qquad (4.37c)$$

where $s(t)$ denotes the transmitted waveform, $y(t)$ denotes the channel output, and $r(t)$ denotes the received waveform. The equalities are in the sense of mean-square convergence [62]. The coordinates s_i, y_i, and r_i are related to $s(t)$, $y(t)$, and $r(t)$ through the expressions

$$s_i = \int s(t)\chi_i(t)\, dt, \qquad (4.38a)$$

$$y_i = \int y(t)\chi_i(t)\, dt, \qquad (4.38b)$$

and

$$r_i = y_i + n_i \qquad (4.38c)$$

with

$$n_i = \int n(t)\chi_i(t)\, dt. \qquad (4.38d)$$

We recall that, given $s(t)$, $y(t)$ is a zero-mean Gaussian random process and hence the y_i are zero-mean Gaussian random variables. Moreover since the channel is linear, the y_i must be of the form

$$y_i = \sum_j h_{ij} s_j . \qquad (4.39)$$

Since the y_i are conditionally zero-mean and Gaussian for any given set of s_j, so also are the h_{ij}.

Equations 4.38 and 4.39 imply that the r_i, which completely specify the received waveform, are related to the s_i, which completely specify the transmitted waveform, and to the n_i, which completely specify the noise waveform, through the expression

$$r_i = \sum_j h_{ij} s_j + n_i. \tag{4.40}$$

If the additive noise is white of power density $N_0/2$, the n_i are statistically independent zero-mean Gaussian random variables with the common variance $N_0/2$. The h_{ij} are, of course the zero-mean Gaussian random variables that specify the channel through their covariance matrix.

A comparison of (4.40) with Fig. 4.5 clearly indicates the nature of our orthogonality constraint. We required that h_{ij} vanish when i differs from j and also that the h_{ii} be statistically independent of each other. We satisfied these requirements by choosing the expansion functions $\chi_i(t)$ appropriately and by considering a restricted class of transmitted waveforms. In particular, we considered waveforms for which all of the nonzero s_j were equal to each other and for which each s_j was nonzero for a single transmitted waveform.

It is no more difficult to determine the optimum receiver for the system of (4.40) than it is for that of Fig. 4.5, but the performance of the more general system is yet to be determined [18–26]. However some generalizations are possible. For example, without relaxing the constraints on the h_{ij}, we can admit more general choices of the s_j in (4.40). This generalization is important when a limited bandwidth is available [6]. Another generalization of Fig. 4.5, which has received some attention, results when the path gains are allowed to be correlated random variables [63–65]. Other extensions that relax the requirement that h_{ij} vanish for i not equal to j seem to have received less attention.

Another extension that has been considered involves the supposition that the path gains possess a mean value [66–72]. Roughly stated, this corresponds to channels that possess a specular component. The analysis of specular channels is similar to that which we consider here although there are, of course, more parameters involved.

The extensions we have described thus far pertain to the system performance that is attainable with an optimum receiver. Another line of investigation has been directed toward the development of suboptimum receivers that are effective in combating intersymbol memory and interference.

Tracking Receivers

The optimum receiver for sequences obviously makes optimum use of any intersymbol memory that may be present, but it may be exceedingly complicated. The optimum receiver for single transmissions, on the other hand, may be much simpler than the sequence receiver, but it makes no use of any available channel memory. One compromise approach, which has been quite successful, is to incorporate a channel-tracking subsystem into a single-transmission receiver [73–81]. The purpose of this subsystem is to " observe " the channel over the entire memory span in an attempt to estimate various channel parameters that enter into the receiver. In some systems a special pilot tone is transmitted to facilitate this estimation, whereas in others the normal information sequence is employed. The first attempt to exploit the channel memory with a tracking receiver was embodied in the Rake system [73].

Equalizing Receivers

The conflict between the optimality and practicality of receivers for sequences and single transmissions is even more pronounced in the presence of intersymbol interference than it is in the presence of memory. This is so because the presence of memory alters the optimality, but not the performance, of the optimum single-transmission receiver, whereas the presence of interference can drastically affect its performance.

The effects of this interference are so severe that, until recently, all fading dispersive communication systems were designed to preclude it. Now, however, some progress has been made in the development of "equalizing" systems that can effectively combat intersymbol interference. These systems are similar to those which exploit memory in that they involve a tracking subsystem that estimates the properties of the channel. However equalizing systems usually require an estimate of the channel impulse response, as contrasted to memory-tracking systems which usually involve only a few channel parameters. Much of the work on equalizing systems has been directed toward channels that are only slightly dispersive in frequency, that it, that are almost time invariant [82–88]. However more attention is now being given to channels, such as HF, that fade more rapidly [89, 90].

REFERENCES

[1] J. M. Wozencraft and I. M. Jacobs, *Principles of Communication Engineering*. New York: Wiley, 1965, Chapters 5–7.
[2] R. G. Gallager, *Information Theory and Reliable Communication*. New York: Wiley, 1968.
[3] R. M. Fano, *Transmission of Information*. New York: Wiley, and the M.I.T. Press, 1961.
[4] R. Ash, *Information Theory*. New York: Interscience, 1965.
[5] G. L. Turin, "Error Probabilities for Binary Symmetric Ideal Reception through Nonselective Slow Fading and Noise." *Proc. IRE*, pp. 1603–1619, September 1958.

[6] J. S. Richters, "Communication Over Fading Dispersive Channels." M.I.T. Research Laboratory of Electronics, *Tech. Rept.*, 464, November 30, 1967.

[7] E. C. Titchmarsh, *The Theory of Functions*. London: Oxford University Press, 1939, p. 403.

[8] "West Ford Issue," *Proc. IEEE*, May 1964.

[9] M. Schwartz, W. R. Bennett, and S. Stein, *Communication Systems and Techniques*. New York: McGraw-Hill, 1966, p. 380.

[10] J. L. Hollis, "An Experimental Equipment to Reduce Teleprinter Errors in the Presence of Multipath," *IRE Trans. Comm. Systems*, pp. 185–188, September 1959.

[11] A. R. Schmidt, "A Frequency Stepping Scheme for Overcoming the Disastrous Effects of Multipath Distortion on High-Frequency FSK Communications Circuits." *IRE Trans. Commun. Systems*, pp. 44–47, March 1960.

[12] P. M. Woodward, *Probability and Information Theory, With Applications To Radar*. London: Pergamon, 1953, Chapter 4.

[13] D. J. Sakrison, *Communication Theory: Transmission of Waveforms and Digital Information*. New York: Wiley, 1968, Chapter 8.

[14] J. M. Wozencraft and I. M. Jacobs, *Principles of Communication Engineering*. New York: Wiley, 1965, Chapter 4.

[15] C. W. Helstrom, *Statistical Theory of Signal Detection*. New York: Pergamon, 1960, Chapter 3.

[16] W. W. Harman, *Principles of the Statistical Theory of Communication*. New York: McGraw-Hill, 1963, Chapters 10 and 11.

[17] W. B. Davenport, Jr. and W. L. Root, *Random Signals and Noise*. New York: McGraw-Hill, 1958, Chapter 14.

[18] J. M. Wozencraft and I. M. Jacobs, *Principles of Communication Engineering*. New York: Wiley, 1965, Chapter 7.

[19] J. C. Hancock and P. A. Wintz, *Signal Detection Theory*. New York: McGraw-Hill, 1966, Chapter 7.

[20] Carl W. Helstrom, *Statistical Theory of Signal Detection*. New York: Pergamon, 1960, Chapter 11.

[21] E. J. Baghdady, Ed., *Lectures on Communication System Theory*. J. M. Wozencraft. New York: McGraw-Hill, 1961, Chapter 12, p. 279.

[22] R. Price, "Optimum Detection of Random Signals in Noise, with Application to Scatter-Multipath Communication, I." *IRE Trans. Inform. Theory*, pp. 125–135, December 1956.

[23] G. L. Turin, "Communication through Noisy, Random-Multipath Channels." *IRE Convention*, pp. 154–166, 1956.

[24] D. Middleton, "On the Detection of Stochastic Signals in Additive Normal Noise—Part I." *IRE Trans. Inform. Theory*, pp. 86–121, June 1957.

[25] T. Kailath, "Correlation Detection of Signals Perturbed by a Random Channel." *IRE Trans. Inform. Theory*, pp. 361–366, June 1960.

[26] P. Bello, "Some Results on the Problem of Discriminating Between Two Gaussian Processes." *IRE Trans. Inform. Theory*, pp. 224–233, October 1961.

[27] A. V. Balakrishnan, Ed., *Communication Theory*. W. L. Root, New York: McGraw-Hill, 1968, Chapter 4, p. 160.

[28] J. Capon, "Hilbert Space Methods for Detection Theory and Pattern Recognition." *IEEE Trans. Inform. Theory*, pp. 247–259, April 1965.

[29] T. T. Kadota, "Optimum Reception of Binary Sure and Gaussian Signals." *Bell System Techn. J.*, pp. 1621–1658, October 1965.

[30] ———, "Optimum Reception of *M*-ary Gaussian Signals in Gaussian Noise." *Bell System Techn. J.*, p. 2187, November 1965.

[31] ——— and L. A. Shepp, "On the Best Finite Set of Linear Observables for Discriminating Two Gaussian Signals." *IEEE Trans. Inform. Theory*, pp. 278–284, April 1967.

[32] J. M. Wozencraft and I. M. Jacobs, *Principles of Communication Engineering*. New York: Wiley, 1965, pp. 528–550.

[33] J. C. Hancock and P. A. Wintz, *Signal Detection Theory*. New York: McGraw-Hill, 1966, pp. 191–198.

[34] M. Schwartz, W. R. Bennett, and S. Stein, *Communication Systems and Techniques*. New York: McGraw-Hill, 1966, Chapters 9–11.

[35] J. N. Pierce, "Theoretical Diversity Improvement in Frequency-Shift Keying." *Proc. IRE*, pp. 903–910, May 1958.

[36] D. G. Brennan, "Linear Diversity Combining Techniques." *Proc. IRE*, pp. 1075–1102, June 1959.

[37] B. B. Barrow, "Error Probabilities for Telegraph Signals Transmitted on a Fading FM Carrier." *Proc. IRE*, pp. 1613–1929, September 1960.

[38] J. N. Pierce, "Theoretical Limitations on Frequency and Time Diversity for Fading Binary Transmissions." *IRE Trans. Commun. Systems*, pp. 186–189, June 1961.

[39] P. M. Hahn, "Theoretical Diversity Improvement in Multiple Frequency Shift Keying." *IRE Trans. Commun. Systems*, pp. 177–184, June 1962.

[40] R. Price, "Error Probabilities for Adaptive Multichannel Reception of Binary Signals." *IEEE Trans. Inform. Theory*, pp. 305–316, September 1962.

[41] W. F. Walker, "The Error Performance of A Class of Binary Communications Systems in Fading and Noise." *IEEE Trans. Commun. Systems*, pp. 28–45, March 1964.

[42] W. C. Lindsey, "Error Probabilities for Rician Fading Multichannel Reception of Binary and N-ary Signals." *IEEE Trans. Inform. Theory*, pp. 339–350, October 1964.

[43] J. Ziv, "Probability of Decoding Error for Random Phase and Rayleigh Fading Channels." *IEEE Trans. Inform. Theory*, pp. 53–61, January 1965.

[44] J. N. Pierce, "Plurality-Count Diversity Combining for Fading M-ary Transmissions." *IEEE Trans. Commun. Technol.*, pp. 529–532, August 1966.

[45] J. M. Wozencraft and I. M. Jacobs, *Principles of Communication Engineering*. New York: Wiley, 1965, p. 546.

[46] M. Schwartz, W. R. Bennett, and S. Stein, *Communication Systems and Techniques*. New York: McGraw-Hill, 1966, p. 590.

[47] G. L. Turin, "The Characteristic Function of Hermitian Quadratic Forms in Complex Normal Variables." *Biometrika*, pp. 199–201, June 1960.

[48] R. G. Gallager, *Information Theory and Reliable Communication*. New York: Wiley, 1968, Chapters 5–8.

[49] R. M. Fano, *Transmission of Information*. New York: Wiley, and M.I.T. Press, 1961, Chapter 9.

[50] C. Cherry, Ed., *Information Theory* (Third London Symposium). P. Elias, "Coding for Two Noisy Channels." New York: Academic, and London: Butterworths, 1956, p. 61.

[51] C. E. Shannon, "Certain Results in Coding Theory for Noisy Channels." *Information and Control*, 1, 625, September 1957.

[52] ———, "Probability of Error for Optimal Codes in a Gaussian Channel." *Bell System Techn. J.*, 611, May 1959.

[53] S. Silver, Ed., *Monograph on Radio Waves and Circuits*. P. Elias, "Coding for Practical Communication Systems," 1963, New York: Elsevier, p. 125.

[54] J. M. Wozencraft and I. M. Jacobs, *Principles of Communication Engineering*. New York: Wiley, 1965, Chapter 6.

[55] Robert G. Gallager, *Information Theory and Reliable Communication*. New York: Wiley, 1968, Chapter 6.

[56] G. D. Forney, *Concatenated Codes*. M.I.T. Press, Cambridge, Mass., 1967.

[57] J. L. Massey, *Threshold Decoding*. M.I.T. Press, Cambridge, Mass., 1963.

[58] E. R. Berlekamp, *Algebraic Coding Theory*. New York: McGraw-Hill, 1968.

[59] W. W. Peterson, *Error Correcting Codes*. New York: Wiley, and M.I.T. Press, 1962.

[60] E. J. Baghdady, Ed., *Lectures on Communication System Theory*. P. Elias, New York: McGraw-Hill, 1961, Chapter 13.

[61] A. Kohlenberg, "An Experimental Comparison of Coding vs. Frequency Diversity for HF Telegraphy Transmission." *IEEE Trans. Commun. Technology*, pp. 532, August 1966.

[62] W. B. Davenport, Jr. and W. L. Root, *Random Signals and Noise*. New York: McGraw-Hill, 1958, pp. 96.

[63] G. L. Turin, "On Optimal Diversity Reception." *IRE Trans. Inform. Theory*, pp. 154–166, July 1961.

[64] J. N. Pierce and S. Stein, "Multiple Diversity with Nonindependent Fading." *Proc. IRE*, pp. 89–104, January 1960.

[65] G. L. Turin, "On Optimal Diversity Reception, II." *IRE Trans. Commun. Systems*, pp. 22–31, March 1962.

[66] W. C. Lindsey, "Asymptotic Performance Characteristics for the Adaptive Coherent Multireceiver and Noncoherent Multireceiver Operating Through the Rician Fading Multichannel." *IEEE Trans. Commun, Electron*. pp. 67–73, January 1964.

[67] W. C. Lindsey, "Error Probabilities for Coherent Receivers in Specular and Random Channels." *IEEE Trans. Inform. Theory*, pp. 147–150, January 1965.

[68] W. C. Lindsey, "Error Probability for Incoherent Diversity Reception." *IEEE Trans. Inform. Theory*, pp. 491–499, October 1965.

[69] W. C. Lindsey, "Error Probabilities for Partially Coherent Diversity Reception." *IEEE Trans. Commun. Technology*, pp. 620–625, October 1966.

[70] I. M. Jacobs, "Probability-of-Error Bounds for Binary Transmission on the Slowly Fading Rician Channel." *IEEE Trans. Inform. Theory*, pp. 431–441, October 1966.

[71] G. D. Hingorani, "Error Rates for a Class of Binary Receivers." *IEEE Trans. Commun. Technology*, pp. 209–215, April 1967.

[72] J. G. Proakis, "On the Probability of Error for Multichannel Reception of Binary Signals." *IEEE Trans. Communic. Technology*, pp. 68–70, February 1968.

[73] R. Price and P. E. Green, "Communication Technique for Multipath Channels." *Proc. IRE*, pp. 555–570, March 1958.

[74] S. M. Sussman, "A Matched Filter Communication System for Multipath Channels." *IRE Trans. Inform. Theory*, pp. 367–373, June 1960.

[75] J. C. Hancock, "Optimum Performance of Self-Adaptive Systems Operating Through a Rayleigh-Fading Medium." *IEEE Trans. Commun. Systems*, pp. 443–453, December 1963.

[76] J. G. Proakis, P. R. Drouilhet, Jr., and R. Price, "Performance of Coherent Detection Systems Using Decision-Directed Channel Measurement." *IEEE Trans. Commun. Systems*, pp. 54–63, March 1964.

[77] G. D. Hingorani and J. C. Hancock, "A Transmitted Reference System for Communication in Random or Unknown Channels." *IEEE Trans. Commun. Technology*, pp. 293–301, September 1965.

[78] R. Esposito, D. Middleton, and J. A. Mullen, "Advantage of Amplitude and Phase Adaptivity in the Detection of Signals Subject to Slow Rayleigh Fading." *IEEE Trans. Inform. Theory*, pp. 473–481, October 1965.

[79] D. R. Bitzer, D. A. Chesler, R. Ivers, and S. Stein, "A Rake System for Tropospheric Scatter." *IEEE Trans. Commun. Technology* pp. 499–506, August 1966.

[80] R. W. Chang, "On Receiver Structures for Channels Having Memory." *IEEE Trans. Inform. Theory*, pp. 463–468, October 1966.

[81] N. J. Bershad, "Optimum Binary FSK for Transmitted Reference Systems Over Rayleigh Fading Channels." *IEEE Trans. Commun. Technology*, pp. 784–790, December 1966.

[82] R. W. Lucky, "Automatic Equalization for Digital Communication," *Bell System Tech. J.*, 547–588, April 1965.

[83] D. A. George, "Matched Filters for Interfering Signals," *IEEE Trans. Inform. Theory*, 153, January 1965.

[84] M. R. Aaron and D. W. Tufts, "Intersymbol Interference and Error Probability." *IEEE Trans. Inform. Theory*, 26–35, January 1966.

[85] R. W. Lucky, "Techniques for Adaptive Equalization of Digital Communication Systems." *Bell System Tech. J.*, 255–286, February 1966.

[86] R. W. Lucky and H. R. Rudin, "Generalized Automatic Equalization for Communication Channels." *Proc. IEEE*, 439, March 1966.

[87] C. W. Niessen and P. R. Drouilhet, "Adaptive Equalizer for Pulse Transmission." 1967 *IEEE* International Conference on Communications.

[88] M. E. Austin, "Decision-Feedback Equalization for Digital Communication Over Dispersive Channels." M.I.T. Research Laboratory of Electronics Tech. Report 461, August 11, 1967.

[89] M. J. DiToro, "A New Method of High-Speed Adaptive Serial Communication Through any Time-Variable and Dispersive Transmission Medium." Conference Record, IEEE Communication Convention, Boulder, Colorado, pp. 763, June 1965.

[90] K. Folkestad, *Ionospheric Radio Communications*. New York: Plenum, pp. 341–358, 1968.

Chapter 5

Error Probability

Our objective is to obtain relatively simple and explicit bounds to the error probability of the fading dispersive communication systems described in Chapter 4. These results are presented in terms of the eigenvalues of the complex correlation function, the average-received-energy–to–noise power-density ratio, the alphabet size, and the information rate. We first obtain results that are valid for any fading dispersive system. We then determine the set of eigenvalues, or equivalent diversity path strengths, that yield the best performance; we conclude that the best system is an equal-strength diversity system. The optimization of this equal-strength system is then presented and its performance evaluated. Finally, we explore the behavior of nonoptimum equal-strength systems and establish bounds that relate the performance of general fading dispersive systems to equal-strength systems.

Roughly stated, we find that the effects of fading and dispersion can be offset by a power increase that varies from about 5 db at low rates to 0 db at high rates. We also find that the optimum energy–to–noise power-density ratio per diversity path increases from about 5 db to infinity as the rate increases from zero to capacity.

The bounds we obtain are rather complex and require considerable manipulation before they admit a simple interpretation. Therefore we digress momentarily and consider the channel for which the only disturbance is additive white Gaussian noise. This digression provides an introduction to the use of error-probability bounds and to the concept of channel reliability. It also serves to introduce some specific results that are used subsequently.

5.1 ADDITIVE WHITE GAUSSIAN NOISE CHANNELS

The system to be considered here is neither dispersive nor fading. The received waveform is, before the addition of the noise, an exact replica of the transmitted waveform: neither its phase nor amplitude are perturbed. It is worth noting that these conditions cannot prevail in systems described by the point-scatterer model of Chapter 2. Even when the communication medium is composed of a single stationary scatterer, that is, is nondispersive, it is still a fading medium in the sense that the scatterer cross section is a random variable.

We now wish to establish bounds on the error probability $P(\varepsilon)$ of the Gaussian channel. We suppose that m-ary orthogonal signaling is employed and that the receiver is designed to minimize the error probability. The exact specification of the waveforms is unimportant; all that matters is that they be orthogonal and of equal energy.

Our supposition leads to a receiver consisting of a set of m branch demodulators that are, in fact, matched filters [1–3]. The outputs of the filters possess the properties discussed in connection with (4.25) and Fig. 4.9. Insofar as these outputs are concerned, the system considered here differs from fading dispersive systems only in the specification of the probability densities of the correct and incorrect branch demodulator outputs. In the present instance, the outputs of the incorrect branches are zero-mean Gaussian random variables with unit variances, whereas the output of the correct branch is a Gaussian random variable with unit variance and mean value $\sqrt{2\alpha}$, where α is the average-received-energy–to–noise ratio per transmission, that is,

$$\alpha = \frac{E_r}{N_0}.$$

The fact that the branch receiver outputs are Gaussian greatly simplifies the reduction of the error-probability bounds given by (4.30). This reduction yields results that can be expressed in terms of the *information rate R*, the *channel capacity C*, and the time duration or *constraint length τ* of the transmission.

The specification of τ serves to fix the time scale of the communication system; that is, τ is defined to be the time allotted to the transmission of a waveform. Since $\log_2 m$ bits of information are conveyed by each transmission, the information rate R of the system is

$$R = \frac{\log_2 m}{\tau} \quad \text{bits/sec.} \tag{5.1a}$$

In the notation of Chapter 4, τ equals $1/r$.

The average received power P of the system is also defined with respect to the constraint length τ; that is,

$$P = \frac{E_r}{\tau} \quad \text{watts,} \tag{5.1b}$$

where E_r is the average received energy due to the transmission of a single waveform. Finally, the capacity C of the system is given by the expression†

$$C = \frac{P}{N_0 \ln 2} = 1.44 \frac{P}{N_0} \qquad \text{bits/sec} \qquad (5.1c)$$

or equivalently

$$C = 1.44 \frac{\alpha}{\tau}. \qquad (5.1d)$$

An upper bound to $P(\varepsilon)$ for the Gaussian channel is derived in Section B of Appendix 2. The result, which differs only in detail from the well-known bounds [4–7], is as follows:

$$P(\varepsilon) \leq K_2 \cdot 2^{-\tau C E} = K_2 \exp{-\alpha E}, \qquad (5.2a)$$

where

$$E = \frac{1}{2} - \frac{R}{C}, \qquad \text{for} \quad 0 \leq \frac{R}{C} \leq \frac{1}{4},$$

$$E = \left(1 - \sqrt{\frac{R}{C}}\right)^2, \qquad \text{for} \quad \frac{1}{4} < \frac{R}{C} \leq 1. \qquad (5.2b)$$

The value of K_2 can be taken to be one; however when α is large, the following values yield sharper bounds to $P(\varepsilon)$:

$$K_2 = \frac{1}{\sqrt{2\pi\alpha}}, \qquad\qquad\qquad\qquad \text{for} \quad 0 \leq \frac{R}{C} \leq \frac{1}{4},$$

$$K_2 = \frac{1}{2\sqrt{\pi\alpha}} \left\{ \frac{1}{1 - \sqrt{R/C}} + \frac{\sqrt{C/R}}{2\sqrt{\pi\alpha}(2\sqrt{R/C} - 1)} \right\}, \qquad \text{for} \quad \frac{1}{4} < \frac{R}{C} \leq 1.$$

$$(5.2c)$$

To a first approximation, the numerical value of the bound to $P(\varepsilon)$ is dominated by the exponential term. This is not to say that the value of the bound is accurately determined by the exponential term. Rather, if the value is to be small, the exponential term must be small; that is, τC or α must be large. Since the exponential term is then much smaller than the right member of (5.2c), we sometimes choose, in the interests of simplicity, to set $K_2 = 1$. An improved, but still simple, estimate of $P(\varepsilon)$ can be obtained by using

† Here, and elsewhere, we use ln to denote natural logarithms.

the asymptotic value of K_2 in the limit of increasing τ with R and C fixed, that is,†

$$K_2 \sim \frac{1}{\sqrt{2\pi\alpha}}, \qquad\qquad 0 \leq \frac{R}{C} \leq \frac{1}{4},$$

$$K_2 \sim \frac{1}{2\sqrt{\pi\alpha}\left(1 - \sqrt{R/C}\right)}, \qquad \frac{1}{4} < \frac{R}{C} \leq 1. \tag{5.3}$$

The upper bound to the error probability is of some interest since it indicates a level of system performance that we can be assured of achieving. Its engineering utility is, however, greatly enhanced by the fact that it is a relatively sharp bound. That is, as shown in Appendix 2, the value of $P(\varepsilon)$ cannot be too much smaller than the value of the upper bound. In particular, the error probability satisfies the inequality

$$P(\varepsilon) \geq K_1 \cdot 2^{-\tau CE}, \tag{5.4a}$$

where τ, C, and E are defined as before and where K_1 does not vary "too rapidly" with τ.

The coefficient K_1 of (5.4a) is more complicated than the corresponding coefficient in the upper bound of (5.2). It is given by the following expression, which is derived in Section B of Appendix 2:

$$K_1 = \frac{1}{4\pi\alpha}\left(1 - \frac{2}{\alpha}\right)^2, \qquad\qquad 0 \leq \frac{R}{C} \leq \frac{1}{4},$$

$$K_1 = \frac{1}{16\pi\alpha}\frac{\sqrt{C/R}}{1 - \sqrt{R/C}}\left[1 - \frac{1}{2\alpha}\left(\frac{C}{R}\right)\right]\left[1 - \frac{1}{2\alpha\left(1 - \sqrt{R/C}\right)^2}\right], \qquad \frac{1}{4} < \frac{R}{C} \leq 1. \tag{5.4b}$$

A less formidable expression is obtained if we suppose that τC tends to infinity for a fixed value of R/C. This asymptotic expression is

$$K_1 \sim \frac{1}{4\pi\alpha}, \qquad\qquad 0 \leq \frac{R}{C} \leq \frac{1}{4},$$

$$K_1 \sim \frac{1}{16\pi\alpha}\frac{\sqrt{C/R}}{\left(1 - \sqrt{R/C}\right)}, \qquad \frac{1}{4} < \frac{R}{C} < 1. \tag{5.5}$$

A comparison of (5.2) through (5.5) shows that the upper and lower bounds to the error probability possess the same exponential dependence upon τCE. Moreover for sufficiently large values of τ, the coefficients of the two bounds do not differ enormously. Of course, for small enough values of τ, or large

† The symbol $\sim$ denotes that the ratio of the two members of the expression approach one.

enough values of the error probability, the "coefficients" may be more important in the computation of $P(\varepsilon)$ than the exponential is. However the exponential term and the asymptotic expressions for the coefficients are very useful in the study of low-error-probability systems

The preceding formulation of the bounds to the error probability emphasizes the time scale of the system through the use of the information rate, power-to-noise ratio, and time-constraint length τ. There is a second formulation that stresses the dependence of the error probability upon energy-to-noise ratios and upon the number v of information bits conveyed per use of the channel. The latter formulation, which is used in the sequel, is obtained by noting that the energy–to–noise-power-density ratio per information bit β satisfies the expression

$$\beta = \frac{\alpha}{v} \tag{5.6a}$$

or

$$\beta = \frac{C}{R} \ln 2, \tag{5.6b}$$

where

$$v = \log_2 m. \dagger \tag{5.6c}$$

Using (5.6) to eliminate R/C from the bounds of (5.2) and (5.4), we obtain the following results:

$$P(\varepsilon) \le K_2 \cdot 2^{-vE_b} \tag{5.7a}$$

and

$$P(\varepsilon) \ge K_1 \cdot 2^{-vE_b}, \tag{5.7b}$$

where E_b, *the reliability per information bit*, is given by the expression

$$E_b = \left[\left(1 - \left(\frac{\beta}{\ln 2}\right)^{1/2}\right)\right]^2 \qquad \text{for} \quad \ln 2 \le \beta \le 4 \ln 2,$$

$$\tag{5.8}$$

$$E_b = \frac{\beta}{2 \ln 2} - 1, \qquad \text{for} \quad \beta \ge 4 \ln 2,$$

and where K_1 and K_2 are determined by (5.4b) and (5.2c). For engineering calculations, this formulation of the bound is usually preferable to that of (5.2) through (5.4), because the efficiency and complexity of a communication system are often best measured by β and v.

---

† Following common usage, we often refer to β as the energy-to-noise ratio per information bit.

We have now completed our discussion of the classical additive white Gaussian noise channel. The analysis has not been particularly difficult because the detailed structure of the transmitted waveforms was irrelevant, the only requirement being that the waveforms be orthogonal to each other. We turn now to the corresponding analysis of fading dispersive channels. This analysis is considerably more involved than that just presented because the detailed structure of the received waveforms, as reflected through the path strengths λ_i, is of paramount concern.

5.2 FADING DISPERSIVE CHANNELS

Our objective is to reduce the bounds of (4.30) to a more manageable form under the supposition that the demodulator-output statistics are given by (4.26) and Fig. 4.9. The central problem in this reduction is to obtain suitable estimates for the three probability distribution functions appearing in those bounds. Since this is purely a mathematical problem, its solution is relegated to Section C of Appendix 2. Since the solution itself is rather complicated, the general form of the result will be presented first and then the properties of its constituent parts will be examined.

The bounds to the error probability can be expressed as

$$P(\varepsilon) \geq K_1 \cdot 2^{-\tau C E} \tag{5.9a}$$

and

$$P(\varepsilon) \leq K_2 \cdot 2^{-\tau C E}. \tag{5.9b}$$

In these expressions τ is the time allotted to the transmission of a waveform and C is the capacity of that classical additive white Gaussian noise channel which possesses the same average power-to-noise ratio as the given fading dispersive channel does, that is,

$$C = \frac{P}{N_0 \ln 2} = \frac{\alpha}{\tau} \frac{1}{\ln 2}. \tag{5.10}$$

The coefficients K_1 and K_2 are discussed in detail subsequently. For the moment it suffices to say that their role is analogous to that of the corresponding coefficients encountered in the discussion of the additive white Gaussian noise channel.

In contrast to the additive white Gaussian noise channel, the function E now depends not only upon R/C but also upon τ and the λ_i. Consequently it is appropriate to regard E as a channel reliability only if we regard the specification of the modulation, and hence the τ and λ_i, as an integral and

unalterable part of the channel. If, as is often the case, the choice of modulation is a design variable, it is more appropriate to regard E as the reliability
of a given communication system. We employ the latter point of view in
the ensuing discussion.

System Reliability

The system reliability function E may be expressed either parametrically
or as the solution of a maximization problem. Also, it can be transformed
into a reliability per information bit as in Section 5.1. Since all of the formu-
lations are of value subsequently, we discuss them now.

Parametric Expressions: The function E may be expressed, parametrically,
in terms of the function

$$\gamma(s) = -\frac{1}{\alpha} \sum_i [\ln (1 - s\alpha\lambda_i) + s \ln (1 + \alpha\lambda_i)], \qquad s \leq 0, \qquad (5.11)$$

where α is the average-received-energy–to–noise ratio and the λ_i are the
average fractional path strengths in the canonical diversity representation of
the system (Fig. 4.5). These parametric expressions possess one of two forms
depending upon whether the information rate exceeds a value, denoted by
Rcrit, called the critical rate of the system. The value of the critical rate is
given by the expression,

$$R\text{crit} = C[\tfrac{1}{2}\gamma'(-\tfrac{1}{2}) - \gamma(-\tfrac{1}{2})]. \qquad (5.12)$$

The parametric expressions for E are as follows:

$$E = -2\gamma(-\tfrac{1}{2}) - \frac{R}{C} \qquad (5.13a)$$

for $0 \leq R \leq R$crit and

$$E = s\gamma'(s) - \gamma(s) \qquad (5.13b)$$

with

$$\frac{R}{C} = (s + 1)\gamma'(s) - \gamma(s) \qquad (5.13c)$$

for Rcrit $\leq R \leq C\gamma'(0)$ or equivalently for values of s in the interval $-\tfrac{1}{2}$ to 0.
In these expressions $\gamma'(s)$ denotes the derivative of $\gamma(s)$ with respect to s.

We now consider the properties of the function E under the supposition
that the energy–to–noise ratio α and the λ_i are fixed. Its general form is
illustrated in Fig. 5.1.

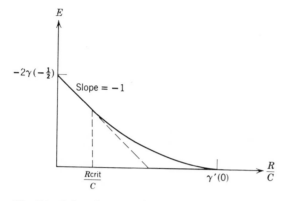

Fig. 5.1 Salient features of the system reliability curve.

Since $-2\gamma(-\frac{1}{2})$ is positive, E is a linear positive monotone-decreasing function of R/C for $0 \le R \le R$crit, and its derivative with respect to R/C in this interval is -1.

For rates greater than the critical rate, the variation of E with R/C is somewhat more complex. However, we can show that, as s increases from $-\frac{1}{2}$ to zero, R increases from Rcrit to $C\gamma'(0)$, E decreases from its value at Rcrit to zero, and the derivative $dE/d(R/C)$ increases from -1 to zero [8]. These conclusions follow from (5.13) in conjunction with the observations that $\gamma(0)$ equals zero and that

$$\frac{dE}{d(R/C)} = \frac{s}{1+s}.$$

Nonparametric Formulation: The definition of the function E may be presented in an alternate form that is of considerable value. To arrive at this form, we first note that, for each value of s between $-\frac{1}{2}$ and 0, the values of E and R/C may be determined by the graphical construction of Fig. 5.2a. That is, the value of $-E$ is given by the $t = 0$ intercept of the straight line tangent to the function $\gamma(t)$ at the point $t = s$, and the value of $-R/C$ is given by the $t = -1$ intercept of this same line. On the other hand, if R is less than Rcrit, the value of E is determined by the graphical construction of Fig. 5.2b. That is, $-E$ equals the $t = 0$ intercept of that straight line which passes through the point $t = -\frac{1}{2}, \gamma(-\frac{1}{2})$ and the point $t = -1, -R/C$.

Now let us consider the line shown in Fig. 5.2c, that is, the straight line that passes through the points $t = -1, -R/C$ and $t = x, \gamma(x)$. We seek to minimize the $t = 0$ intercept of this line by varying x over the interval $-\frac{1}{2} \le x \le 0$. Clearly this intercept is minimized, that is, made as negative as possible, when the straight line is tangent to the $\gamma(t)$ curve. Moreover as is

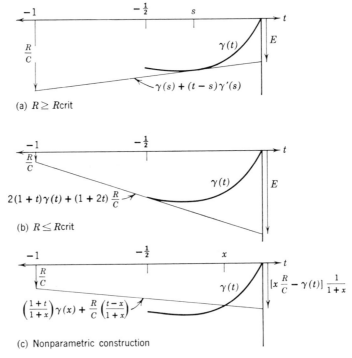

(a) $R \ge R\text{crit}$

(b) $R \le R\text{crit}$

(c) Nonparametric construction

Fig. 5.2   Graphical construction for determining E as a function of R/C.

easy to verify, $\gamma(t)$ is convex so there is at most one point of tangency. Of course, if R/C is small enough, no tangency point exists in the interval $-\frac{1}{2} \le x \le 0$, and the minimum occurs when x equals $-\frac{1}{2}$. In any event, the value of the minimum equals the value of $-E$ for the given value of R/C. Consequently the value of E is given by the expression

$$E = \max \left[-\frac{\gamma(x)}{1+x} + \left(\frac{x}{1+x}\right)\frac{R}{C} \right],$$

where the maximization is over x in the interval $-\frac{1}{2} \le x \le 0$.

Alternately we may replace $-x/(1+x)$ with ρ, define the function

$$E(\rho) = \frac{1}{\alpha} \sum_i \left[(1+\rho)\ln\left(1 + \frac{\alpha\rho\lambda_i}{1+\rho}\right) - \rho \ln(1 + \alpha\lambda_i) \right], \qquad (5.14a)$$

that is,

$$E(\rho) = -(1+\rho)\gamma\left(\frac{-\rho}{1+\rho}\right), \qquad (5.14b)$$

and reformulate the expression for E as follows:

$$E = \max_{0 \leq \rho \leq 1} \left[E(\rho) - \rho \frac{R}{C} \right]. \qquad (5.14c)$$

This latter formulation is analogous to that sometimes employed in random coding theorems [9].

The formulation of (5.14) provides an easy means of obtaining upper and lower bounds to the reliability E. In fact, it is obvious that

$$E \leq \max_{0 \leq \rho \leq 1} \left[E_U(\rho) - \rho \frac{R}{C} \right] \qquad (5.15a)$$

and

$$E \geq \max_{0 \leq \rho \leq 1} \left[E_L(\rho) - \rho \frac{R}{C} \right], \qquad (5.15b)$$

where $E_U(\rho)$ and $E_L(\rho)$ are, respectively, any upper and lower bounds to $E(\rho)$. These inequalities are used subsequently to determine the values of the λ_i that maximize E and to establish some useful estimates for E.

Reliability per Information Bit: As noted in Section 5.1, it is frequently convenient to express the error bounds in terms of β, the average-received-energy–to–noise ratio per information bit, and v, the number of information bits transmitted per use of the channel. These parameters are related to the information rate R, constraint length τ, and average-received-energy–to–noise ratio α by (5.6), for example,

$$v = \log_2 m \qquad (5.16a)$$

and

$$\beta = \frac{\alpha}{v}. \qquad (5.16b)$$

Introducing these quantities into (5.9) and (5.14). we obtain

$$P(\varepsilon) \geq K_1 \cdot 2^{-vE_b}, \qquad (5.17a)$$

$$P(\varepsilon) \leq K_2 \cdot 2^{-vE_b}, \qquad (5.17b)$$

where

$$E_b = \frac{C}{R} E$$

or equivalently

$$E_b = \max_{0 \leq \rho \leq 1} \left[\frac{\beta E(\rho)}{\ln 2} - \rho \right]. \qquad (5.17c)$$

The subscript b is added to E to indicate that the exponent is in "units" of inverse information bits.

Just as we are prone to characterize the additive white Gaussian noise channel by its reliability, so also are we prone to characterize fading dispersive channels by the exponents E or E_b of (5.14) and (5.17c). To justify this course of action, we must consider the character of the coefficients K_1 and K_2. That we now do.

Coefficients

It is proved in Appendix 2 (A2.79a) that K_2 may be set equal to 2. It is also shown there that K_1 may be set equal to $\frac{1}{64} \exp -\sqrt{\alpha/|s|}$ if R exceeds Rcrit and equal to $\frac{1}{64} \exp -\sqrt{2\alpha}$ otherwise. Thus the error probability satisfies the double inequality

$$\left[\tfrac{1}{64} \exp - \left(\frac{\alpha}{|s|} \right)^{1/2} \right] \cdot 2^{-\tau CE} \le P(\varepsilon) \le 2 \cdot 2^{-\tau CE}, \qquad (5.18)$$

where $|s|$ is determined by (5.13c) if R exceeds Rcrit and is set equal to $\frac{1}{2}$ otherwise.

Equation 5.18 clearly indicates that the error probability depends exponentially upon the quantity τCE. More precisely, for fixed values of R/C and of s we have

$$\lim_{\tau C \to \infty} \left[\frac{\log_2 P(\varepsilon)}{\tau C} \right] \to - \lim_{\tau C \to \infty} E. \qquad (5.19)$$

This situation is usually described by saying that the rightmost and leftmost members of (5.18) are "exponentially correct."

The exponential dominance that we have established thus far may be exceedingly weak. That is, the upper and lower bounds of (5.18) may differ markedly for relatively large values of τC or equivalently of α. Alternately the value of α required to achieve a particular numerical value of the upper bound may be appreciably greater than the value of α required to attain the same value with the lower bound. This state of affairs is disagreeable because the error-probability bounds are often used to determine bounds to the energy–to–noise ratio that is required for a given error probability.

Fortunately the results of (5.18) can be improved substantially when α, or τC, is moderately large, that is, when the error probability is sufficiently small. Specifically it is shown in Appendix 2 (A2.51) that for any fixed value of s,

$$\lim_{\alpha \to \infty} \alpha K_1 \sim \frac{1}{4|s| \, |1 + s|2\pi\gamma''(s)} \ge \frac{1}{4\pi\alpha}; \qquad (5.20)$$

that is, for sufficiently large values of α, the coefficient in the lower bound to the error probability varies no more rapidly than inversely with α rather than exponentially with the square root of α. Moreover as shown in Appendix 2 (A2.79), the upper-bound coefficient satisfies the inequalities

$$K_2 \leq \frac{1}{\sqrt{\pi\alpha\gamma''(-\frac{1}{2})}}\left(\frac{\eta_c}{\eta_c - 1}\right)^2, \qquad \text{for } R \leq R\text{crit}, \qquad (5.21a)$$

and

$$K_2 \leq \frac{1}{|s|\sqrt{2\pi\alpha\gamma''(s)}}\left(\frac{1 - |s|}{1 - 2|s|}\right)\left(\frac{\eta}{\eta - 1}\right)^2, \qquad \text{for } R\text{crit} \leq R \leq C\gamma'(0), \tag{5.21b}$$

where

$$\eta_c = 2\alpha\gamma''(-\tfrac{1}{2})\left(\frac{1}{2} + \frac{1}{\alpha\lambda_o}\right)^2, \tag{5.22a}$$

$$\eta = \alpha\gamma''(s)\left(|s| + \frac{1}{\alpha\lambda_o}\right)^2 \tag{5.22b}$$

and where λ_o is the maximum value of the λ_i. If there is more than one positive eigenvalue, (5.21a) and (5.21b) yield results that vary inversely with $\sqrt{\alpha\gamma''(-\frac{1}{2})}$ and $\sqrt{\alpha\gamma''(s)}$, respectively. Thus for large values of α, the ratio of the upper and lower bounds to $P(\varepsilon)$ provided by (5.20) and (5.21) vary as $\sqrt{\alpha/\gamma''(s)}$ rather than exponentially with $\sqrt{\alpha}$ as the bounds provided by (5.18) do.

The expressions for K_1 and K_2 can be further sharpened and simplified if we impose restrictions on the eigenvalues λ_i. One restriction of particular interest is that all of the positive eigenvalues be equal; that is,

$$\lambda_i = \frac{1}{D}, \qquad i = 1, \ldots, D. \tag{5.23a}$$

Subject to this restriction, the values of K_1 and K_2 satisfy the following inequalities:

$$K_1 \geq \frac{0.2}{2\pi}\frac{1 + \alpha|s|/D}{(1 + D)(1 + \alpha/D)}, \tag{5.23b}$$

$$K_2 \leq \frac{1}{\sqrt{\pi\alpha}}\left(\frac{D}{\alpha}\right)^{1/2}\left(1 + \frac{\alpha}{2D}\right)\left(\frac{2D}{2D - 1}\right)^2, \qquad \text{for } R \leq R\text{crit},$$

$$K_2 \leq \frac{1}{|s|\sqrt{2\pi\alpha}}\left(\frac{D}{\alpha}\right)^{1/2}\frac{1 - |s|}{1 - 2|s|}\left(1 + \frac{\alpha|s|}{D}\right)\left(\frac{D}{D - 1}\right)^2, \qquad \text{for } R\text{crit} < R \leq C\gamma'(0). \tag{5.23c}$$

The derivation of these inequalities is presented in Appendix 2, [(A2.53) and (A2.82)]. Another refinement, which we do not consider, is possible for rates less than Rcrit [10].

For the most part we focus upon the exponential part of the bounds rather than upon the coefficients. Thus rather than stating explicit bounds, we often employ approximations of the form

$$P(\varepsilon) \approx 2^{-\nu E_b},$$

where the symbol $\approx$ denotes "approximately equal to."

At this juncture the basic bounds to the error probability of fading dispersive channels are established. It is clear, from (5.14), that the application of these bounds involves the fractional path strengths λ_i. In a true diversity system these strengths are readily determined. However for a scattering channel they are the eigenvalues of the complex correlation function and can be determined often only with great difficulty. Even when the eigenvalues can be determined, it is seldom possible to express E, or even $E(\rho)$, in a closed form.

Of course the bounds may be evaluated for any given system by resorting to numerical computation. Moreover very efficient computational procedures are available for some classes of correlation functions [11–13]. Thus the prospect of a numerical analysis need not be depressing. The difficulty is rather that we prefer to know the dependence of the system's performance upon the various design parameters, and such dependencies are difficult to determine by numerical analysis. Therefore we seek to eliminate some of the complexity from the expression for the reliability by examining special classes of systems.

Specifically we first determine the structure of the canonical diversity system that yields the maximum value of E for any given values of R/C and α. We find that the best system is one in which all of the positive eigenvalues are equal to each other and their number is determined by the values of α and R/C. This conclusion leads us to a study of equal-strength diversity systems. Bounds to the performance of nonoptimum systems are then established in terms of "equivalent" optimum systems.

5.3 THE OPTIMUM FADING DISPERSIVE CHANNEL

The value of the system reliability E depends upon R/C, α, and the λ_i. We now seek to maximize E with respect to the λ_i, for fixed values of α and R/C, subject to some constraints on the λ_i. The supposition that α and R/C are fixed is quite reasonable. It is satisfied, for example, if the propagation medium, receiver front-end temperature, transmitter power, information rate, and waveform transmission rate (r or τ^{-1}) are fixed.

Ideally we would seek to maximize E by appropriately choosing the basic modulator complex envelope $u(t)$ described in Chapter 4. That is, we would maximize E with respect to the λ_i, subject to the constraint that they be eigenvalues of the complex correlation function $R(t, \tau)$ defined by (4.4) for some $u(t)$. Unfortunately this maximization appears to be exceedingly difficult to perform. Therefore we consider the related but simpler problem of maximizing E with respect to the λ_i, subject only to the constraints that the λ_i be nonnegative and that their sum be unity. This is equivalent to assuming that the designer is free to specify both the modulation and the channel scattering function. Although this may appear to be rather unrealistic, we find that the results obtained are quite useful in the design of systems.

One final preliminary remark is in order. The system reliability E will be maximized with respect to the λ_i rather than the error probability minimized. This decision is, in part, necessary because the set of λ_i that minimizes the true value of $P(\varepsilon)$ is not known. However the firm lower bound to $P(\varepsilon)$ and also the weak asymptotic lower bound depend on the λ_i only through the reliability E; thus by maximizing E, we are minimizing those bounds to the error probability.

Formulation

The mathematical problem we face is that of maximizing the value of E with respect to the λ_i, subject to the constraints that

$$\lambda_i \geq 0 \qquad (5.24a)$$

and that

$$\sum_i \lambda_i = 1. \qquad (5.24b)$$

The maximization is to be performed for fixed values of R/C and α.

The parametric expressions for E are not well suited to the task we face, but the formulation of (5.14c) is quite useful because the order of maximization with respect to ρ and the λ_i can be interchanged. In particular, the maximum value E° of E may be expressed either as

$$E^\circ = \max_{\{\lambda_i\}} \left\{ \max_{0 \leq \rho \leq 1} \left[E(\rho) - \rho \frac{R}{C} \right] \right\} \qquad (5.25a)$$

or equivalently as

$$E^\circ = \max_{0 \leq \rho \leq 1} \left\{ \left[\max_{\{\lambda_i\}} E(\rho) \right] - \rho \frac{R}{C} \right\}. \qquad (5.25b)$$

Thus our problem is reduced to that of maximizing $E(\rho)$ with respect to the λ_i for given values of ρ, R/C, and α.

As a first step in the maximization, we express the quantity $E(\rho)$, of (5.14), as

$$E(\rho) = \sum_i \lambda_i f_\rho(\alpha \lambda_i), \qquad (5.26)$$

where the function $f_\rho(x)$ is defined by the expression

$$f_\rho(x) = \rho \left[\frac{(1 + \rho)}{\rho x} \ln \left(1 + \frac{x\rho}{1 + \rho} \right) - \frac{1}{x} \ln (1 + x) \right]. \qquad (5.27)$$

We next show that, for any fixed value of ρ, $f_\rho(x)$ possesses a single maximum as a function of x. We also establish some other properties of $f_\rho(x)$ which are required subsequently.

Properties of $f_\rho(x)$: The function $f_\rho(x)$ has the general form illustrated in Fig. 5.3. For every value of ρ, it is a nonnegative function for all positive values of x, and its value tends to zero as x tends to either zero or infinity. Moreover $f_\rho(x)$ is a concave function of x for sufficiently small positive values of x, and a convex function for sufficiently large values of x. These conclusions follow easily from (5.27); those which we now state are not quite so obvious.

For every fixed ρ, the function $f_\rho(x)$ is concave for all values of x less than some value y_0 and is convex for all values of x exceeding y_0. It attains its maximum at a point $\alpha_p{}^\circ$ whose value is the unique positive solution of the equation

$$f_\rho(\alpha_p{}^\circ) = \left(\frac{\rho}{1 + \rho} \right) \frac{\alpha_p{}^\circ}{(1 + \alpha_p{}^\circ)(1 + \alpha_p{}^\circ \rho/(1 + \rho))}. \qquad (5.28)$$

The proofs of these assertions are presented in the following paragraphs. The reader who is uninterested in mathematical detail may omit them and proceed to the optimality of equal-strength diversity systems.

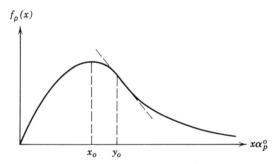

Fig. 5.3   Important features of $f_\rho(x)$.

Since $f_\rho(x)$ is nonnegative and its value is zero at both end points, it must possess at least one maximum. Moreover if the function possesses any local minima, they must be preceded and followed by local maxima. That is, if there are $n/2$ local maxima and $n/2 - 1$ local minima, then there must exist numbers

$$x_1 < x_2 < \cdots < x_{n-1} \tag{5.29}$$

such that

$$f'_\rho(x_i) = 0, \qquad i = 1, \ldots, n-1, \tag{5.30a}$$

$$f''_\rho(x_i) < 0, \qquad i = \text{even}, \tag{5.30b}$$

and

$$f''_\rho(x_i) > 0, \qquad i = \text{odd}, \tag{5.30c}$$

where $f'_\rho(x)$ and $f''_\rho(x)$ are, respectively, the first and second derivatives of $f_\rho(x)$ with respect to x for fixed ρ.

The derivatives $f'_\rho(x)$ and $f''_\rho(x)$ satisfy the relation

$$xf''_\rho(x) + 2f'_\rho(x) = \frac{\rho^2}{(1 + x)^2(1 + \rho + \rho x)^2}\left(\frac{1 + \rho}{\rho} - x^2\right). \tag{5.31}$$

Thus (5.30a) and (5.30b) can be satisfied only if x_i exceeds $\sqrt{(1 + \rho)/\rho}$ for i even, whereas (5.30a) and (5.30c) can be satisfied only if the x_i are less than $\sqrt{(1 + \rho)/\rho}$ for i odd. However x_2 exceeds x_1, so n must equal one. Consequently $f_\rho(x)$ possesses a single maximum which occurs at the value x_1 and which we denote by $\alpha_\rho{}^\circ$. The value of $\alpha_\rho{}^\circ$ can be determined from the expression $f'_\rho(x) = 0$ that yields (5.28).

We now show that $f_\rho(x)$ is first concave and then convex. It can be verified by direct computation that

$$\frac{d}{dx}\left[x^3 f''(x)\right] = +2\rho x^2\left[\frac{-1}{(1 + x)^3} + \frac{\delta^2}{(1 + \delta x)^3}\right], \tag{5.32}$$

where $\delta = \rho/(1 + \rho)$. The right member of this expression is negative if

$$x < (1 - \delta^{2/3})(\delta^{2/3} - \delta)^{-1} = y_1, \tag{5.33}$$

and zero or positive otherwise. That is, $x^3 f''_\rho(x)$ decreases as x increases from 0 to y_1. Since $f''_\rho(0)$ is negative, this implies that $f''_\rho(x)$ is negative; hence $f_\rho(x)$ is concave for all positive values of x less than y_1. For values of x greater than y_1, $x^3 f''_\rho(x)$ increases continuously; thus if $f''_\rho(x)$ becomes positive, it remains positive. Since $f''_\rho(x)$ is positive for sufficiently large values of x, we conclude that $f_\rho(x)$ is first concave and then convex. The value y_0 of x for which $f_\rho(x)$ changes from concave to convex must, of course, exceed $\alpha_\rho{}^\circ$.

Optimality of Equal-strength Diversity Systems: The maximization of the function $E(\rho)$ of (5.26), with respect to the λ_i, is now reduced to a trivial problem. In fact,

$$f_\rho(\alpha\lambda_i) \leq f_\rho(\alpha_p{}^o), \qquad i = 1, \ldots . \tag{5.34a}$$

Hence

$$E(\rho) \leq \sum_i \lambda_i f_\rho(\alpha_p{}^o) = f_\rho(\alpha_p{}^o) \sum_i \lambda_i = f_\rho(\alpha_p{}^o), \tag{5.34b}$$

where the rightmost equality follows from the constraint that the λ_i sum to unity. Thus $E(\rho)$ cannot exceed $f_\rho(\alpha_p{}^o)$, where $\alpha_p{}^o$ is determined as a function of ρ from (5.28).

On the other hand, if precisely $\alpha/\alpha_p{}^o$ of the λ_i are positive and if each of them equals $\alpha_p{}^o/\alpha$, $E(\rho)$ equals $f_\rho(\alpha_p{}^o)$. Thus we conclude that $E(\rho)$ is maximized by setting $\alpha/\alpha_p{}^o$ of the λ_i equal to $\alpha_p{}^o/\alpha$ and setting the remaining λ_i equal to zero. Since the fading dispersive systems we are considering can be visualized as diversity systems with relative path strengths λ_i, we may paraphrase our result as follows: *The most reliable fading dispersive system is an equal-strength diversity system.*

We have now established the general form of the optimum system. We next determine the value of the positive λ_i in terms of ρ and determine the resulting maximum reliability E^o.

The Optimized System

The optimum system is most easily specified in terms of $\alpha_p{}^o$, which we call the *energy–to–noise ratio per diversity path.* Since the value of α is presumed to be fixed, the required value of $\alpha_p{}^o$ can be achieved only by adjusting D, the number of equal positive eigenvalues, to equal $\alpha/\alpha_p{}^o$. A difficulty arises because D must be an integer, whereas $\alpha/\alpha_p{}^o$ may assume noninteger values.

We can account for the integer constraint on the value of D in the optimization, but if $\alpha/\alpha_p{}^o$ exceeds unity, the refinement of the results is often too small to warrant their additional complication. Moreover we can ensure, by a perturbation of α, that the value of $\alpha/\alpha_p{}^o$ is an integer. Thus the discrepancy introduced by ignoring the requirement that D be an integer is not very important when $\alpha/\alpha_p{}^o$ exceeds unity.

When $\alpha/\alpha_p{}^o$ is less than unity, the situation changes markedly. In this instance it is unsatisfactory to suppose that D equals $\alpha/\alpha_p{}^o$; we must instead determine the maximum value of $f_\rho(\alpha\lambda_i)$ subject to the integer constraint. The determination is accomplished by noting that $f_\rho(\alpha\lambda_i)$ is less than $f_\rho(\alpha)$ when α is less than $\alpha_p{}^o$. Thus the maximum value of $E(\rho)$ is $f_\rho(\alpha)$, and this value is attained when only one of the λ_i is positive. That is, if $\alpha_p{}^o$ exceeds α, D should be set equal to unity, or equivalently a single diversity path should be employed. This is the first of many encounters with threshold phenomena in fading dispersive channels. It is worth noting that this threshold phenomenon is always encountered as ρ is decreased for any fixed value of α.

We have now established that $E(\rho)$ is maximized, for a given value of ρ, when the λ_i correspond to a D-fold equal-strength diversity system. The optimum value D° of D is determined by the relation

$$D^\circ = \frac{\alpha}{\alpha_p{}^\circ}, \qquad \text{if} \quad \alpha > \alpha_p{}^\circ, \tag{5.35a}$$

and

$$D^\circ = 1, \qquad \text{otherwise.} \tag{5.35b}$$

The value of $\alpha_p{}^\circ$, as a function of ρ, is determined by (5.28), and the resulting maximum value of $E(\rho)$ is $f_\rho(\alpha_p{}^\circ)$ if α exceeds $\alpha_p{}^\circ$ and is $f(\alpha)$ otherwise. Our next tasks are to determine the values of $\alpha_p{}^\circ$ and of E° as functions of R/C, or of β, rather than as functions of ρ.

Determination of $\alpha_p{}^\circ$: By virtue of (5.14) and (5.34), in conjunction with the comment following (5.35), the optimum value E° of E is given by the expression

$$E^\circ = \max_{0 \le \rho \le 1} \left[f_\rho(\alpha_p{}^\circ) - \rho \frac{R}{C} \right] \tag{5.36}$$

when α exceeds $\alpha_p{}^\circ$ where $\alpha_p{}^\circ$ is determined, as a function of ρ, by (5.28). In the following paragraphs we suppose that α does exceed $\alpha_p{}^\circ$ and determine the dependence of $\alpha_p{}^\circ$ upon R/C.

To proceed, we evaluate the derivative of the quantity $f_\rho(\alpha_p{}^\circ) - \rho R/C$ with respect to ρ for a fixed value of R/C. This evaluation is simplified by recalling that the partial derivative of the function $f_\rho(x)$ with respect to x is zero at the point $x = \alpha_p{}^\circ$ for any fixed value of ρ, that is,

$$\left. \frac{\partial f_\rho(x)}{\partial x} \right|_\rho = 0 \qquad \text{for} \quad x = \alpha_p{}^\circ, \tag{5.37}$$

where the symbol $|_\rho$ denotes that ρ is held fixed in the partial differentiation.

By the chain rule of partial differentiation

$$\frac{d}{d\rho} \left[f_\rho(\alpha_p{}^\circ) - \rho \frac{R}{C} \right] = \left. \frac{\partial f_\rho(\alpha_p)}{\partial \alpha_p{}^\circ} \right|_\rho \frac{d\alpha_p{}^\circ}{d\rho} + \left. \frac{\partial f_\rho(\alpha_p{}^\circ)}{\partial \rho} \right|_{\alpha_p{}^\circ} - \frac{R}{C}, \tag{5.38}$$

and hence by (5.37)

$$\frac{d}{d\rho} \left[f_\rho(\alpha_p{}^\circ) - \rho \frac{R}{C} \right] = \left. \frac{\partial f_\rho(\alpha_p{}^\circ)}{\partial \rho} \right|_{\alpha_p{}^\circ} - \frac{R}{C}. \tag{5.39}$$

The required partial derivative of $f_\rho(x)$ can be evaluated from (5.27) to obtain

$$\frac{d}{d\rho}\left[f_\rho(\alpha_p{}^\circ) - \rho\frac{R}{C}\right] = \frac{1}{\alpha_p{}^\circ}\ln\left(1 - \frac{\alpha_p{}^\circ}{1 + \alpha_p{}^\circ}\frac{1}{1 + \rho}\right) + \frac{1}{1 + \rho(1 + \alpha_p{}^\circ)} - \frac{R}{C}.$$

(5.40)

The value of $\alpha_p{}^\circ$ is determined, as a function of ρ, by (5.28), that is,

$$f_\rho(\alpha_p{}^\circ) = \frac{\rho}{1 + \rho}\frac{\alpha_p{}^\circ}{(1 + \alpha_p{}^\circ)[1 + \alpha_p{}^\circ\rho/(1 + \rho)]},$$

(5.41)

and the function $f_\rho(x)$ is defined by (5.27), that is,

$$f_\rho(x) = \rho\left[\frac{1 + \rho}{\rho x}\ln\left(1 + x\frac{\rho}{1 + \rho}\right) - \frac{1}{x}\ln(1 + x)\right].$$

(5.42)

It is complicated, but not difficult, to show that the right member of (5.40) is a decreasing function of ρ for $0 \le \rho \le 1$. Therefore the function $[f_\rho(\alpha_p{}^\circ) - \rho R/C]$ possesses a single maximum in the interval $0 \le \rho \le 1$. It can also be shown that the right member of (5.40) is positive when ρ equals zero provided that R/C is less than one. Hence the maximum occurs at the point $\rho = 1$ if the right member of (5.40) is then positive; otherwise the maximum occurs at that (unique) value of ρ for which the right member of (5.40) equals zero.

In summary the value of E° is given by the equation

$$E^\circ = f_1(\alpha_p{}^\circ) - \frac{R}{C}$$

(5.43a)

provided that

$$\frac{R}{C} \le \frac{1}{\alpha_p{}^\circ}\ln\left(1 - \frac{1}{2}\frac{\alpha_p{}^\circ}{1 + \alpha_p{}^\circ}\right) + \frac{1}{2 + \alpha_p{}^\circ} = \frac{R\text{crit}}{C},$$

(5.43b)

where the value of $\alpha_p{}^\circ$ in (5.43) is determined from (5.41) with ρ set equal to unity. On the other hand, if R exceeds Rcrit, the value of E° is given by the expression

$$E^\circ = f_\rho(\alpha_p{}^\circ) - \rho\frac{R}{C},$$

(5.44)

where ρ and $\alpha_p{}^\circ$ are now the unique solutions of the equations

$$\frac{R}{C} = \frac{1}{\alpha_p{}^\circ}\ln\left(1 - \frac{\alpha_p{}^\circ}{1 + \alpha_p{}^\circ}\frac{1}{1 + \rho}\right) + \frac{1}{1 + \rho(1 + \alpha_p{}^\circ)},$$

(5.45a)

$$f_\rho(\alpha_p{}^\circ) = \frac{\rho}{1 + \rho}\frac{\alpha_p{}^\circ}{(1 + \alpha_p{}^\circ)[1 + \alpha_p{}^\circ\rho/(1 + \rho)]}$$

(5.45b)

with $0 \le \alpha_p{}^\circ$ and $0 \le \rho \le 1$, where $f_\rho(x)$ is defined by (5.42).

Equations 5.43 through 5.45 are not particularly appealing, but they can be solved numerically for each value of R/C. This has been done to obtain the results presented in Fig. 5.4.

The value of $\alpha_p{}^\circ$ is plotted as a function of R/C in Fig. 5.4a and as a function of β, the energy–to–noise ratio per information bit, in Fig. 5.4b. As illustrated in Fig. 5.4a, the value of $\alpha_p{}^\circ$ is roughly equal to 3 for values of R less than about $0.04C$. This result was first obtained in a somewhat different context [14].

For rates greater than the critical rate $\alpha_p{}^\circ$ is a monotone increasing function of R/C. Thus for a fixed value of α, the optimal number of (equal-strength) diversity paths decreases as we seek to operate at greater rates. Equivalently $\alpha_p{}^\circ$ increases monotonically as β is decreased below the critical value of about 12 db.

It is clear from Fig. 5.4 that the value of $\alpha_p{}^\circ$ increases without limit as we consider larger values of R/C, or smaller values of β. Consequently for any given value of α, a point is eventually reached at which the value of $\alpha_p{}^\circ$ is equal to α, that is, where the optimum value D° of D is unity. Any further increase in rate requires a number of diversity paths less than unity. This is the threshold situation referred to in conjunction with (5.35), and as noted then, the value of D° under these circumstances is unity. That is, the value

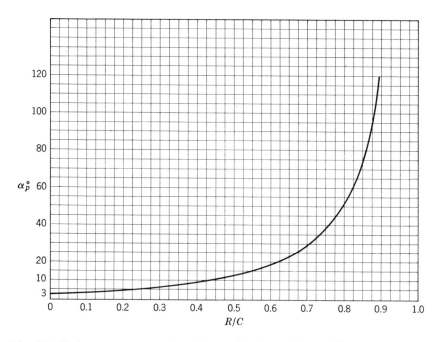

Fig. 5.4 Optimum energy-to-noise ratio per diversity path $\alpha_p{}^\circ$: (a) versus normalized information rate R/C; (b) versus the energy-to-noise ratio per information bit β, in decibels.

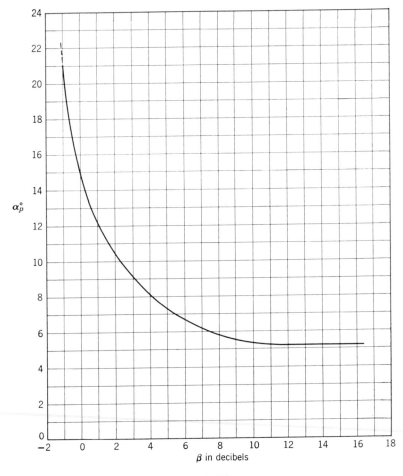

Fig. 5.4(b)

of D° is α/α_p°, with α_p° determined by Fig. 5.4, provided that the value thus determined exceeds unity. Otherwise, the value of D° is unity.

Determination of E°: Now that the values of α_p° and of D° are established, we determine the resulting value of E°. For sufficiently large values of α, the value of D° is α/α_p° and E° is determined by (5.43) to (5.45). It is convenient to consider first this limiting behavior of E° and then to examine the threshold effects that occur with finite values of α.

The value of E° is plotted in Fig. 5.5 as a function of R/C under the assumption that, for each value of R/C, α exceeds the value of α_p° as determined from Fig. 5.4. The reliability function of the additive white Gaussian

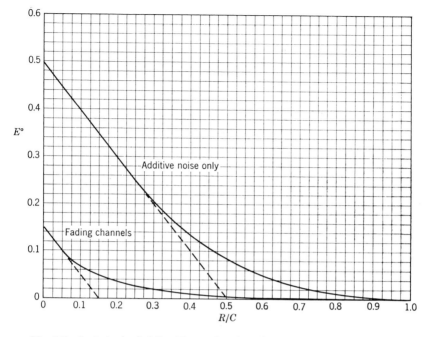

Fig. 5.5   Optimized reliability E° as a function of normalized information rate.

noise channel is shown for purposes of comparison. The salient characteristics of the two curves are as follows.

First, the reliability of the (optimum) fading dispersive channel is positive for all values of R less than C, the capacity of the additive white Gaussian noise channel. Thus an arbitrarily small error probability can be achieved with either of the two channels provided that R is less than C and that τ (or α) is sufficiently large. It can also be shown that the error probabilities cannot be made arbitrarily small for any rate greater than C. Consequently the capacity of the optimized fading dispersive channel equals the capacity of that additive white Gaussian noise channel which has the same average-power-to-noise ratio. This capacity without the attendant error-probability results has been obtained previously [15].

Although the capacities of the additive white Gaussian noise channel and fading dispersive channels are equal, they have little else in common. In particular the reliabilities of the two channels differ markedly for rates less than C, and the additive white Gaussian noise channel performs better than the fading dispersive channel does for the same system complexity.

For example, let us suppose that C and R are, respectively, 40,000 and 5,000 bits/sec and that m equals 32. The value of τ must be 10^{-3} sec; hence

τC equals 40. Thus for an additive white Gaussian noise channel

$$P(\varepsilon) \approx 2^{-40(3/8)} = 2^{-15} \approx 3 \times 10^{-5}. \qquad (5.46)$$

However for the fading dispersive channel, the value of $\alpha_p{}^\circ$ is about 4, the associated value of E° is 0.055; hence

$$P(\varepsilon) \approx 2^{-(40)(0.055)} = 2^{-2.2} \approx 0.2. \qquad (5.47)$$

In the example just considered, the difference between the error probabilities for the two channels is rather large. However the system error probabilities frequently change drastically when the system power levels are changed only slightly. Therefore it is also of interest to compare the relative power levels required by the two channels for a given error probability.

One useful formulation for this comparison is discussed in conjunction with (5.17). It involves β, the energy–to–noise ratio per information bit; v, the number of information bits per transmitted waveform; and E_b, the reliability per information bit. These quantities are related to those of Fig. 5.5 by the equations

$$v = \log_2 m = R\tau, \qquad (5.48a)$$

$$\beta = \frac{\alpha}{v} = \frac{C\tau}{v} \ln 2 = \left(\frac{C}{R}\right) \ln 2, \qquad (5.48b)$$

and

$$E_b{}^\circ = \frac{C}{R} E^\circ. \qquad (5.48c)$$

The superscript o has been added to E_b to emphasize that it has been optimized. The resulting estimate for the error probability is

$$P(\varepsilon) \approx 2^{-vE_b{}^\circ}. \qquad (5.49)$$

The "exponent" $E_b{}^\circ$ has been plotted in Fig. 5.6 as a function of β for both the additive Gaussian noise channel and the optimum fading dispersive channel. The curves clearly indicate the larger value of β required by the fading dispersive channel to achieve a desired error probability for a given value of v. In fact, for large values of β, the energy–to–noise ratio per information bit required by the fading dispersive channel is approximately 3.3 times the value required by the additive Gaussian noise channel. That is, at large values of β *the fading dispersive channel requires about 5 db more power than the additive Gaussian noise channel does* for the same performance and the same value of v. At lower values of β, or at higher rates, the curves indicate that little additional power is required by the fading dispersive channel. In fact, as capacity is approached, no additional power is needed.

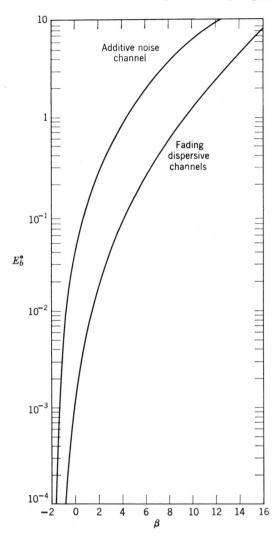

Fig. 5.6 Optimized reliability E_b^o, as a function of the energy-to-noise ratio per information bit β in decibels.

The engineering implications of the preceding results are rather interesting. The low rate, or high β, performance that is readily achievable with an additive Gaussian noise channel is more difficult, or costly, to achieve with a fading dispersive channel. On the other hand, the two channels have similar theoretical performance characteristics for smaller values of β, but systems employing conventional communication techniques are incapable of operating at these smaller values. The difficulty arises because the value of v required for a reasonable value of the error probability becomes prohibitively large as we seek to decrease β.

It is worth noting that the graphical results of Figs. 5.5 and 5.6 may be expressed analytically for rates less than, or β greater than, the critical value. Specifically, if R does not exceed Rcrit, the optimum energy–to–noise ratio per diversity path is 3, and the resulting reliability is

$$E^\circ = 0.15 - \frac{R}{C}. \tag{5.50a}$$

Equivalently

$$E_b{}^\circ = 0.216\beta - 1 \tag{5.50b}$$

for $\beta \geq \beta$crit $= 12$ db.

For rates between the critical rate and capacity, no simple general analytical description of E° is available, but for rates sufficiently close to C, relatively simple expressions can be obtained. These expressions are of considerable value since the graphs of Figs. 5.5 and 5.6 are useless for rates close to the channel capacity.

To obtain the desired expressions, we first determine, from (5.45b), that the parameters ρ and $\alpha_p{}^\circ$ are related as

$$\rho \sim \frac{\ln (\alpha_p{}^\circ)^2}{(\alpha_p{}^\circ)^2}$$

for sufficiently small values of ρ or equivalently for sufficiently large values of $\alpha_p{}^\circ$. Upon introducing this expression into (5.44) and (5.45), we obtain the following parametric expressions among E°, R/C, and $\alpha_p{}^\circ$:

$$E^\circ \sim 2 \frac{(\ln \alpha_p{}^\circ)^2}{(\alpha_p{}^\circ)^3}, \qquad \text{as} \quad \alpha_p{}^\circ \to \infty, \tag{5.51a}$$

$$1 - \frac{R}{C} \sim 3 \frac{\ln \alpha_p{}^\circ}{\alpha_p{}^\circ}, \qquad \text{as} \quad \alpha_p{}^\circ \to \infty. \tag{5.51b}$$

Also

$$E_b{}^\circ \sim 2 \frac{(\ln \alpha_p{}^\circ)^2}{(\alpha_p{}^\circ)^3}, \tag{5.52a}$$

where

$$\beta \sim (\ln 2)\left(1 + 3 \frac{\ln \alpha_p{}^\circ}{\alpha_p{}^\circ}\right). \tag{5.52b}$$

In these expressions the symbol $\sim$ denotes that the ratio of the two sides of the expression approaches unity with increasing $\alpha_p{}^\circ$.

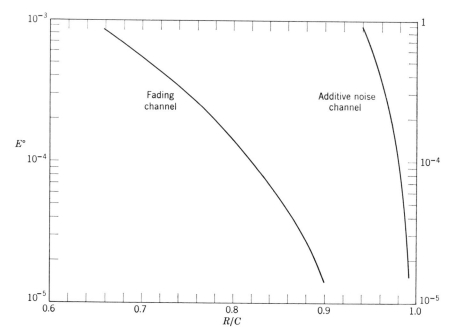

Fig. 5.7 Asymptotic behavior of E^o for information rates approaching capacity.

Equations 5.51 are plotted as one curve in Fig. 5.7 with the corresponding results for an additive Gaussian noise channel. It is clear from the figure that the reliability of the fading dispersive channel approaches zero much more rapidly than that of the additive Gaussian noise channel.

Thus far we have supposed that α is exceedingly large. The extension of our results to account for fixed finite values of α is straightforward. In fact, we recall that D^o equals $\alpha/\alpha_p^{\ o}$ provided that $\alpha/\alpha_p^{\ o}$ exceeds unity; otherwise D^o equals unity; for example, if the specified value of α is less than 3, D^o equals unity for all values of the rate, and the resulting value of E^o may then be computed from the expressions [(5.43) through (5.45a)] for E^o with $\alpha_p^{\ o}$ replaced by α. Note that $f_p(\alpha)$ must be evaluated from (5.42) rather than from (5.45b). The results of these computations for various values of α are presented in Fig. 5.8. Each curve in the figure is labeled with the value of α for which the computation is performed. The significance of the curves labeled with values of α greater than 3 is explained below.

If the specified value of α is greater than 3, D^o equals $\alpha/3$ for values of rate less than the critical rate. More generally, D^o equals $\alpha/\alpha_p^{\ o}$ for all rates such that $\alpha/\alpha_p^{\ o}$ exceeds unity. However for each value of α greater than 3, there is a threshold rate at which $\alpha/\alpha_p^{\ o}$, and hence D^o, equals unity. For rates less than the threshold, the optimized E versus R/C curve is identical to that

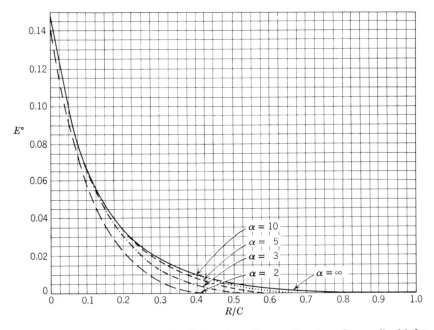

Fig. 5.8 Optimized reliability E° for finite values of α as a function of normalized information rate.

derived under the supposition that α is exceedingly large. For rates greater than the threshold, the maximum value of E for the specified value of α is determined by (5.43) through (5.45a) with α_p° replaced by α. Thus we obtain a family of curves as shown in Fig. 5.8.

We have now shown that the optimized system is an equal-strength diversity system with a particular value of α_p, the energy–to–noise ratio per diversity path. Subsequently we approach the problem from a different vantage point and investigate the performance of equal-strength diversity systems as a function of α_p. First, however, we briefly relate the performance of coded systems to uncoded systems. The discussion may be omitted without loss of continuity.

Coded Systems

In this section we illustrate how our results can be applied to coded systems and demonstrate that the performance of a large orthogonal waveform alphabet can be approached by using digital coding in conjunction with a relatively small alphabet of modulation waveforms. Our objectives are to ease the transition from the results of this monograph to those of coding theory and to suggest that there are practical techniques for approaching the performance of large orthogonal alphabets (say 2^{100}) with relative small modulator waveform alphabets (say 16 waveforms). Since comprehensive discussions of these techniques are available, we content ourselves with a very abbreviated discussion [16–20].

According to (4.31a), a code exists whose (word) error probability is no larger than

$$\overline{P(\varepsilon)} \leq 2^{-(N/r)E_c}, \tag{5.53a}$$

where r is the channel waveform transmission rate, N is the number of waveforms contained in a code word, and

$$E_c = \max_{0 \leq \rho \leq 1} [rE_o(\rho) - \rho R] \tag{5.53b}$$

with $E_o(\rho)$ given by (4.35).

As noted in Chapter 4, the evaluation of $E_o(\rho)$ is quite difficult except when ρ equals unity. However we can at least obtain a bound to $\overline{P(\varepsilon)}$ with that choice of ρ. Specifically

$$E_c \geq rE_o(1) - R. \tag{5.54}$$

Moreover it is not difficult to show that E_c, in fact, equals the right member of this expression when R is sufficiently small. In the interests of simplicity we limit the discussion to the bound of (5.54).

After some manipulation, (5.53) and (5.54) in conjunction with (4.36) yield

$$\overline{P(\varepsilon)} \leq 2^{-N[E_o(1)-R/r]}, \tag{5.55a}$$

where

$$E_o(1) = -\log_2 \left[\frac{1}{m} + \left(1 - \frac{1}{m} \right) \exp -\alpha E(1) \right] \tag{5.55b}$$

and where $E(1)$ is the quantity involved in the earlier results of this chapter, that is,

$$E(1) = \frac{1}{\alpha} \sum_i \left[2 \ln \left(1 + \frac{\alpha \lambda_i}{2} \right) - \ln (1 + \alpha \lambda_i) \right]. \tag{5.56}$$

As before, the λ_i in this expression are the eigenvalues of the complex correlation function associated with the basic complex envelope and α is the average-energy–to–noise ratio associated with the reception of the basic modulator waveform.

We now want to investigate the dependence of (5.55) on the system parameters. To facilitate comparison with our uncoded results, the parameters we employ are β, the energy–to–noise ratio per information bit; K, the number of information bits per code word; m, the number of waveforms in the basic modulator waveform set; and a new parameter k,

$$k = \frac{mN}{K},\tag{5.57}$$

which is called the number of degrees of freedom per information bit. The primary significance of k, and the source of its name, is that the required transmission bandwidth is approximately equal to kR. The remaining parameters, which we suppress, are N, which is determined by (5.57); r, which equals kR/m; and α, the energy–to–noise ratio per modulator waveform, which equals $m\beta/k$.

Upon introducing the above parameters into (5.55), we obtain

$$\overline{P(\varepsilon)} \le 2^{-K[(\eta\beta/\ln 2)E(1)-1]},\tag{5.58a}$$

where

$$E(1) = \frac{k}{m\beta} \sum_i \left[2\ln\left(1 + \frac{m\beta\lambda_i}{2k}\right) - \ln\left(1 + \frac{m\beta\lambda_i}{k}\right) \right]\tag{5.58b}$$

and

$$\eta = \frac{-k}{m\beta E(1)} \ln\left[\frac{1}{m} + \left(1 - \frac{1}{m}\right) \exp -\frac{m\beta E(1)}{k} \right].\tag{5.58c}$$

It is instructive to compare (5.58) with the error probability of an N-fold ($N = Kk/m$) explicit diversity system employing the same basic modulator envelope and the same values of K and β. According to the discussion of Section 4.4, the use of N-fold explicit diversity causes each eigenvalue λ_i of the original modulator envelope to be replaced by N eigenvalues each of value λ_i/N. Upon introducing the latter set of eigenvalues into (5.14) and then underbounding E_b of (5.17) by

$$E_b \ge \frac{\beta E(1)}{\ln 2} - 1,\tag{5.59}$$

we obtain the following bound to the error probability of the explicit diversity system

$$P(\varepsilon) \le 2^{-K[\beta E(1)/\ln 2 - 1]},\tag{5.60}$$

where $E(1)$ is given by (5.58b).

Equations 5.58 and 5.60 indicate that the bounds to the two systems are identical when η equals unity. More generally, the bound to the error probability of the coded system behaves as the bound to an explicit diversity system in which the energy–to–noise ratio per information bit is only $\eta\beta$ instead of β. Moreover the bound of (5.60) is, in fact, exponentially correct when β exceeds a critical value and the bound of (5.58a) is the best result that can be obtained from (5.53) when β is sufficiently large. Thus, for sufficiently large values of β, η is an appropriate measure of the performance of the coded system relative to the explicit diversity system. Since the value of η lies in the interval 0 to 1, we view it as an efficiency.

We now want to determine the behavior of η as k, or m, are varied for fixed values of β and K. Although a variation in either k or m usually alters the value of $E(1)$ if β is to remain constant, we initially suppose that $E(1)$ is fixed. The coupling between $E(1)$, k, and m is considered subsequently.

In Fig. 5.9, η is plotted as a function of $m\beta E(1)/k$ for various values of m. The figure indicates that η essentially attains its limiting value of $1 - 1/m$ as soon as $m\beta E(1)/k$ becomes less than about unity. Since $E(1)$ cannot exceed 0.15, we conclude that the limiting value of η is attained, for all practical purposes, when $m\beta/k$, or equivalently α, is less than about 5.

It is fortunate that $m\beta/k$ need not be too small for η to approach its limiting value. This is so because $m\beta/k$ is the energy–to–noise ratio per modulator waveform, and as it decreases $E(1)$ eventually begins to decrease. Thus it is desirable to operate with a value of $m\beta/k$ that is large enough for $E(1)$ to be near its maximum value and yet small enough for η to be approximately equal to $1 - 1/m$. For example, if the system appears to be nondispersive, that is, if there is only one positive eigenvalue, and if $m\beta/k$ equals 3, then $E(1)$ equals 0.15, the value for an optimized equal-strength diversity system. Hence the system performs almost as well as an optimized equal-strength diversity system with an orthogonal alphabet of 2^K waveforms and with an energy–to–noise ratio per information bit of $1 - 1/m$ times that of the given system. Of course, as the system becomes more dispersive, the value of $m\beta/k$ must increase if $E(1)$ is to maintain its maximum value. Consequently k begins to decrease. Thus the independent maximization of η and $E(1)$ becomes impossible as the system becomes more dispersive.

Another approach, which eliminates one of the parameters, is to maximize η with respect to m for a fixed value of β/k, again assuming that $E(1)$ can be held fixed as m is varied. This maximization has been performed numerically to obtain the results shown in Fig. 5.10. The values of m that yield the maximum value of η for several value of $\beta E(1)/k$ are indicated on the curve. It is again clear that $\beta E(1)/k$ need not be too small, nor m too large, in order to approach an η of unity. This is important because it implies that k, and hence the bandwidth, need not be too large for a given value of β.

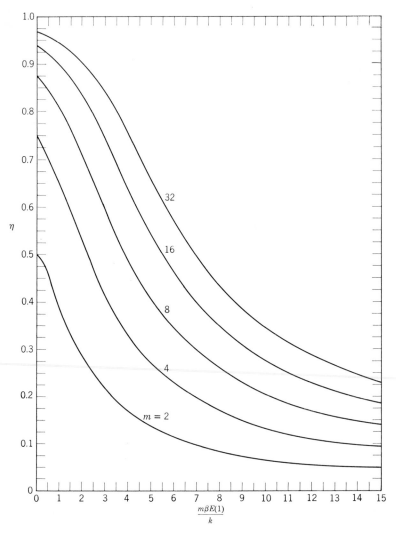

Fig. 5.9 Efficiency η of a coded system as a function of $m\beta E(1)/k$ for various waveform alphabet sizes.

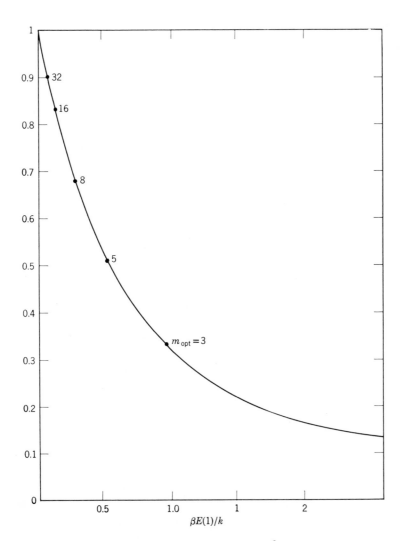

Fig. 5.10 Efficiency η of a coded system as a function of $\beta E(1)/k$ when m is optimized.

5.4 EQUAL-STRENGTH DIVERSITY SYSTEMS

In this section we suppose that precisely D of the eigenvalues are positive and that each of these equals D^{-1}. The corresponding energy–to–noise ratio per diversity path is denoted by α_p. Clearly

$$\alpha_p = \frac{\alpha}{D},$$

where α is the average-received-energy–to–noise ratio resulting from a use of the channel. One of our objectives is to determine the variation in the performance of such a system as the value of α_p is perturbed from $\alpha_p{}^\circ$, the optimum value.

The bounds to the error probability of a D-fold equal-strength diversity system can be determined from (5.9), (5.14), (5.26), and (5.27); that is,

$$K_1 \cdot 2^{-\tau CE} \le P(\varepsilon) \le K_2 \cdot 2^{-\tau CE}, \tag{5.61}$$

where K_1 and K_2 are the coefficients of (5.23), where

$$E = \max_{0 \le \rho \le 1} \left[f_\rho(\alpha_p) - \rho\,\frac{R}{C} \right], \tag{5.62}$$

and where the function $f_\rho(x)$ is defined by the expression

$$f_\rho(x) = \frac{1+\rho}{x} \ln\left(1 + \frac{\rho x}{1+\rho}\right) - \frac{\rho}{x} \ln(1+x). \tag{5.63}$$

The values of E resulting from this expression are presented in Fig. 5.11 and Table 5.1 for several values of α_p.

Table 5.1 **System Reliability, as a Function of R/C, for Equal-Strength Diversity Systems with $\alpha_p = 2, 3, 4, 5, 10, 20, 50,$ and 100**

				R/C				
α_p	0	0.025	0.05	0.075	0.1	0.2	0.3	0.4
2	0.15	0.12	0.095	0.074	0.057	0.021	0.006	0.00016
3	0.15	0.12	0.095	0.08	0.065	0.029	0.012	0.004
4	0.15	0.12	0.098	0.08	0.065	0.033	0.015	0.006
5	0.14	0.11	0.094	0.078	0.065	0.034	0.017	0.0075
10	0.12	0.094	0.075	0.063	0.053	0.03	0.017	0.0095
20	0.088	0.065	0.052	0.044	0.037	0.022	0.013	0.0077
50	0.052	0.035	0.028	0.023	0.019	0.012	0.0075	0.0044
100	0.032	0.02	0.016	0.013	0.011	0.0067	0.0041	0.0026

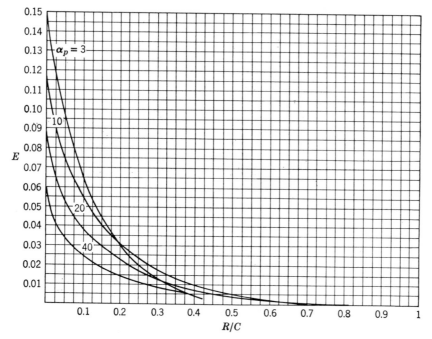

Fig. 5.11 Reliability of equal-strength diversity system for energy-to-noise ratios per diversity path of 3, 10, 20, and 40.

We subsequently investigate the detailed variations of E with α_p at low rates. First, however, we examine the limiting behavior of E as α_p becomes either exceedingly large or exceedingly small. The variation for intermediate values of α_p can be obtained from (5.62) and (5.63) by numerical analysis.

Inadequate Diversity

We determine here the limiting behavior of the error-probability exponent τCE as τC, and hence α, tends to infinity for fixed values of D and of R/C. We find that the value of this limit is

$$\lim_{\tau C \to \infty} \tau CE \to \frac{D}{\ln 2}\left(\frac{R}{C} - 1 - \ln \frac{R}{C}\right). \tag{5.64}$$

This is in marked contrast to the behavior of an optimized system wherein E remains fixed as τC tends to infinity. The comparison provides a testimonial to the importance of signal design in fading dispersive systems.

The limit of (5.64) may be derived as follows. It is easy to verify that the quantity to be maximized in (5.62) possesses a unique maximum in the interval $0 \le \rho \le 1$. For any fixed value of R/C and a sufficiently large value of α_p, this maximum is attained when the value of ρ is determined from the expression

$$\frac{R}{C} = \frac{1}{\alpha_p} \ln \left(1 + \frac{\rho}{1 + \rho} \alpha_p \right) - \frac{1}{\alpha_p} \ln (1 + \alpha_p) + \frac{1}{1 + (1 + \alpha_p)\rho}, \quad (5.65a)$$

where

$$\alpha_p = \frac{\alpha}{D}. \quad (5.65b)$$

Or, in the limit of increasing α_p,

$$\frac{R}{C} = \frac{1}{1 + \alpha_p \rho}. \quad (5.65c)$$

Solving this expression for ρ, introducing the result into (5.62), multiplying by τC, and taking the limit we obtain (5.64).

The limit of (5.64) indicates that the error-probability exponent cannot be made arbitrarily large by merely increasing τ for fixed values of R, C, and D. Of course, the limit we take here implies that the transmitter alphabet size m is increasing exponentially with τC. That is, R/C is presumed to be fixed and

$$m = 2^{\tau R},$$

hence τR must increase linearly with τC and m must increase exponentially with τC. From this point of view the limit may appear to be unrealistic. However the limit of (5.64) is also the exponent that results when β, the energy–to–noise ratio per information bit, is fixed and α, the total energy–to–noise ratio, is increased for a fixed value of D. The latter situation is of considerable engineering interest.

The underlying cause of the performance degradation we observe can be explained as follows. Each equivalent diversity path affords the receiver an independent noisy "sample" of the transmitted waveform. Some of the "samples" are well above the noise level, whereas others are far below it. By increasing the number of diversity paths, we can increase the probability of attaining a total received energy that is not substantially below the average energy. Thus if the energy–to–noise ratio per diversity path is sufficiently large, the performance of the system can be improved by reducing the ratio and increasing the number of paths. However as we now see, this reduction of α_p should not be carried too far.

Excessive Diversity

We now consider the limiting extreme in which α_p tends to zero for a fixed value of R/C. We conclude that the entire error-probability exponent

τCE tends to zero no matter how large τ or C may be. Roughly stated, this result shows that efficient operation cannot be achieved if the available energy–to–noise ratio α is spread "too thin" among the various diversity paths. This spreading is deleterious because of the square-law suppression attendant to the optimum branch demodulators.

To establish the desired result, we note that E attains its maximum value when R equals zero. Hence, upon setting ρ equal to unity and R equal to zero in (5.62), we obtain

$$E \le \frac{2}{\alpha_p} \ln \left(1 + \frac{\alpha_p}{2} \right) - \frac{1}{\alpha_p} \ln (1 + \alpha_p), \tag{5.66}$$

or as α_p tends to zero,

$$E \le \tfrac{1}{4}\alpha_p \to 0. \tag{5.67}$$

More generally, as D tends to infinity for any fixed value of τC (no matter how large), the error-probability exponent τCE tends to zero. Precisely

$$\lim_{D \to \infty} \tau CE \le \lim_{D \to \infty} \left[\left(\frac{\tau C}{2} \right)^2 \frac{\ln 2}{D} \right] = 0. \tag{5.68}$$

Since τCE tends to zero with fixed τC and increasing D, whereas it tends to a positive constant with fixed D and increasing τC we conclude that too much diversity is even worse than too little diversity is.

Low Rate Results

The detailed dependence of E on α_p must, in general, be determined numerically for each value of R/C. However a simple expression does exist for all sufficiently small rates. In particular, it follows from (5.62) that

$$E = f_1(\alpha_p) - \frac{R}{C} \tag{5.69}$$

for all values of R less than the critical rate Rcrit. Thus it is convenient to measure the dependence of E on α_p, for "low" rates, by the quantity $f_1(\alpha_p)$.†
Equivalently we may use the normalized measure μ_0, defined as

$$\mu_0 = \frac{f_1(\alpha_p)}{f_1(\alpha_p^{\circ})} \approx \frac{f_1(\alpha_p)}{0.15}. \tag{5.70}$$

The maximum value of μ_0 is unity, and this value is attained when α_p equals 3.

† The critical rate is itself a function of α_p. Hence the use of the undefined "low."

The quantity μ_o may be interpreted as a relative power (or capacity) efficiency of an equal-strength diversity system. Specifically

$$\tau CE = \tau(\mu_o C)\left(0.15 - \frac{R}{\mu_o C}\right). \tag{5.71}$$

Thus at low rates, a system with given values of R, C, τ, and μ_o performs as though it is that optimum system with the same values of τ and R as the given system but with a capacity equal to μ_o times C.

Alternately we may express the error exponent as

$$\tau CE = vE_b = v(0.216\mu_o\beta - 1), \tag{5.72}$$

where β is the energy–to–noise ratio per information bit and v is the number of information bits conveyed per transmission. Thus μ_o also may be interpreted as an efficiency with respect to the energy–to–noise ratio per information bit. That is, a system with given values of β, v, and μ_o possesses the same error exponent as does that optimum system with the same value of v but with an energy–to–noise ratio per information bit of μ_o times β.

The value of μ_o, as a function of α_p, can be determined from (5.63) in conjunction with (5.70). The resulting function is plotted in Fig. 5.12. The figure clearly demonstrates that the efficiency is relatively insensitive to the energy–to–noise ratio per diversity path. In fact μ_o exceeds 0.75 for energy–to–noise ratios per diversity path ranging from 1 to 10. On the other hand, μ_o does tend to zero if α_p is either exceedingly small or exceedingly large.

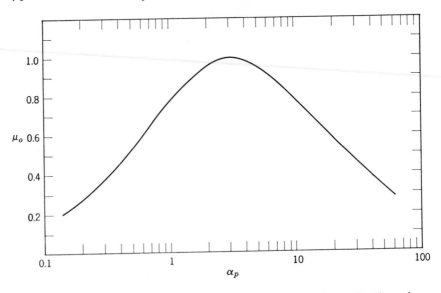

Fig. 5.12   Low rate efficiency μ_o as a function of energy-to-noise ratio per diversity path α_p.

At this juncture we have completed our discussion of equal-strength diversity systems. Our next task is to determine some simple bounds to the reliability of any given fading dispersive system in terms of an equivalent equal-strength system. These bounds underlie our subsequent discussion, in Chapter 6, of the signal design problem.

5.5 RELIABILITY BOUNDS

According to (5.15) upper and lower bounds to the system reliability can be obtained by upper and lower bounding the quantity $E(\rho)$. Our present objective is to obtain bounds of the form $\varepsilon f_\rho(\alpha_p)$, where $f_\rho(x)$ is the function defined by (5.63). We seek such bounds because they enable us to bound the reliability of a given system by the reliability of a related equal-strength diversity system. We first establish this relation and then derive expressions for ε and α_p.

Let us suppose that $E(\rho)$ satisfies the inequality

$$E(\rho) \geq \varepsilon f_\rho(\alpha_p), \tag{5.73a}$$

where ε and α_p are not functions of ρ. By virtue of (5.15), the system reliability then satisfies the expression

$$E \geq \max_{0 \leq \rho \leq 1} \left[\varepsilon f_\rho(\alpha_p) - \rho \frac{R}{C} \right] \tag{5.73b}$$

or

$$\frac{E}{\varepsilon} \geq \max_{0 \leq \rho \leq 1} \left[f_\rho(\alpha_p) - \rho \frac{R}{\varepsilon C} \right]. \tag{5.73c}$$

The right member of (5.73c) is just the reliability of an equal-strength diversity system operating at an energy–to–noise ratio per diversity path of α_p. We take the rate of this fictitious equal-strength system to be R, the rate of the original system, and take its power–to–noise ratio, or "additive Gaussian noise capacity," to be ε times that of the original channel. Thus we conclude that the reliability of the given system is at least as large as ε times the reliability of an equal-strength system operating at the same rate, but with an energy–to–noise ratio per diversity path of α_p and a total power–to–noise ratio of only ε times that of the given system. Since the complete error-probability exponent is $\tau C E$, we also conclude that the error-probability exponent of a system for which (5.73a) is valid is at least as large as that of the equal-strength diversity system whose capacity is εC and whose energy–to–noise ratio per diversity path is α_p.

The preceding discussion suggests that ε be regarded as an estimate of the efficiency of a system relative to that equal-strength diversity system which is associated with the given value of α_p. The utility of this point of view is enhanced subsequently when we find that the value of α_p can be optimized so that ε provides an estimate of the efficiency relative to an optimized equal-strength diversity system.

We can also interpret ε as a measure of the reduction in β that can be realized by replacing a given system by an equal-strength diversity system. This interpretation follows upon recalling that the error-probability exponent can be expressed as vE_b, where

$$E_b = \max_{0 \le \rho \le 1} \left[\frac{\beta E(\rho)}{\ln 2} - \rho \right], \tag{5.74}$$

or by virtue of (5.73a)

$$E_b \ge \max_{0 \le \rho \le 1} \left[\beta \varepsilon \frac{f_\rho(\alpha_p)}{\ln 2} - \rho \right]. \tag{5.75}$$

Since $f_\rho(\alpha_p)$ is just $E(\rho)$ for an equal-strength diversity system, we conclude that the value of E_b for the given system is at least as large as the value of E_b for an equal-strength diversity system operating at an energy–to–noise ratio per diversity path of α_p and an energy–to–noise ratio per information bit of ε times the ratio for the given system. Thus we are led to view $\varepsilon\beta$ as a lower bound to the effective energy–to–noise ratio per information bit. Similarly α_p may be regarded as an estimate of the "effective" energy–to–noise ratio per diversity path, and $\varepsilon\alpha/\alpha_p$, which equals D, may be regarded as an estimate of the "effective" number of diversity paths.

It is a simple matter to extend the preceding results to obtain upper bounds to the system reliability. For example, the validity of the inequality

$$E(\rho) \le \varepsilon f_\rho(\alpha_p) \tag{5.76}$$

for a given system implies that its reliability per information bit E_b is no larger than that of the equal-strength diversity system employing an energy–to–noise ratio per diversity branch of α_p and an energy–to–noise ratio per information bit of ε times that of the given system.

With the significance of the bounds for $E(\rho)$ established, we next establish the bounds themselves. The bounds that follow provide a reasonable compromise between sharpness and complexity. They are used repeatedly in Chapter 6.

Lower Bound

We show here that the inequality

$$E(\rho) \ge \varepsilon f_\rho(\alpha_p) \tag{5.77a}$$

with

$$\varepsilon = \sqrt{bD} \qquad (5.77b)$$

and

$$\alpha_p = \alpha\left(\frac{b}{D}\right)^{1/2} \qquad (5.77c)$$

is valid for all values of ρ in the interval $0 \leq \rho \leq 1$ and for all values of D that do not exceed b^3/d^2. The quantities b, d, and $f_\rho(\cdot)$ are given by the expressions

$$b = \sum_i \lambda_i^2, \qquad (5.78a)$$

$$d = \sum_i \lambda_i^3, \qquad (5.78b)$$

$$f_\rho(x) = \frac{1 + \rho}{x} \ln\left(1 + x\frac{\rho}{1 + \rho}\right) - \frac{\rho}{x} \ln(1 + x), \qquad (5.78c)$$

where the λ_i are the fractional path strengths in the diversity representation of the given system.

To establish the validity of (5.77), we first introduce the function $W(\rho)$ defined as

$$W(\rho) = \frac{-1}{1 + \rho}[\varepsilon f_\rho(\alpha_p) - E(\rho)], \qquad (5.79a)$$

or by virtue of (5.26)

$$W(\rho) = \frac{-1}{1 + \rho}\left[\varepsilon f_\rho(\alpha_p) - \sum_i \lambda_i f_\rho(\alpha\lambda_i)\right]. \qquad (5.79b)$$

We now show that $W(\rho)$ is nonnegative, and hence that (5.77a) is valid, for all values of D which do not exceed b^3/d^2.

It is not difficult to verify that

$$\frac{(1 + \rho)^2}{\alpha}\frac{\partial^2 W(\rho)}{\partial \rho^2} = +\frac{2(1 + \rho)}{\alpha}\frac{\partial W(\rho)}{\partial \rho} + \frac{\alpha_p \varepsilon}{\alpha[1 + \rho(1 + \alpha_p)]^2}$$

$$- \sum_i \frac{\lambda_i^2}{[1 + \rho(1 + \alpha\lambda_i)]^2}, \qquad (5.80)$$

or since $[1 + \rho(1 + y)]^{-2}$ is a convex function of y [21],

$$1(+\rho)^2\frac{\partial^2 W(\rho)}{\partial \rho^2} \leq 2(1 + \rho)\frac{\partial W(\rho)}{\partial \rho} + \frac{\alpha_p \varepsilon}{[1 + \rho(1 + \alpha_p)]^2} - \frac{\alpha b}{[1 + \rho(1 + \alpha d/b)]^2}.$$

Thus if D does not exceed b^3/d^2, α_p exceeds $d\alpha/b$ and

$$(1 + \rho)\frac{\partial^2 W(\rho)}{\partial\rho^2} \leq -2\frac{\partial W(\rho)}{\partial\rho}. \tag{5.81}$$

Equation 5.81 implies that, for positive values of ρ, every stationary point of $W(\rho)$ is a maximum. Thus the minimum value of $W(\rho)$ in the interval $0 \leq \rho \leq 1$ occurs at one of the end points, and it suffices to prove that the end-point values are nonnegative.

It is easy to verify that $W(\rho)$ vanishes when ρ equals zero. To show that $W(\rho)$ is nonnegative when ρ is one, we note that

$$W(1) = -\frac{1}{2}\left[\varepsilon f_1(\alpha_p) - \alpha \sum_i \lambda_i{}^2 B(\alpha\lambda_i)\right],$$

where

$$B(x) = \frac{1}{x^2}\ln\left[1 + \frac{x^2}{4(1 + x)}\right] = \frac{1}{x}f_1(x).$$

We further conclude, after some calculation, that $B(x)$ is convex for non-negative values of x. Consequently

$$W(1) \geq -\frac{1}{2}\left[\varepsilon f_1(\alpha_p) - \frac{b^2}{d}f_1\left(\frac{\alpha d}{b}\right)\right],$$

or if ε and α_p are defined by (5.77),

$$W(1) \geq \frac{1}{2}b\alpha\left[\frac{b}{\alpha d}f_1\left(\frac{\alpha d}{b}\right) - \frac{1}{\alpha}\sqrt{\frac{D}{b}}f_1\left(\alpha\sqrt{\frac{b}{D}}\right)\right].$$

Finally we note, from Fig. 5.3, that $(1/x)f_1(x)$ is the slope of the straight line passing through the origin and the point $x, f_1(x)$. Thus $(1/x)f_1(x)$ is a decreasing function of x, and if D does not exceed b^3/d^2,

$$\frac{1}{\alpha}\sqrt{\frac{D}{b}}f_1\left(\alpha\sqrt{\frac{b}{D}}\right) \leq \frac{b}{\alpha d}f_1\left(\frac{\alpha d}{b}\right);$$

thus $W(1) \geq 0$. Therefore, as asserted, $W(\rho)$ is nonnegative in the interval $0 \leq \rho \leq 1$.

We can lend more substance to (5.77) by considering two examples. Let us first suppose that the given system possesses precisely N equal positive eigenvalues. The values of b and d are then, respectively, N^{-1} and N^{-2}, and the upper limit to D is N. Upon setting D equal to N, α_p becomes α/N whereas ε equals one. That is, the inequality of (5.77a) reduces to an equality for equal-strength diversity systems.

As a second example, consider a system possessing N_1 eigenvalues equal to λ_o and N_2 eigenvalues equal to $(1 - N_1\lambda_o)N_2^{-1}$. The values of ε and α_p, as determined from (5.77b) and (5.77c), with D set equal to b^3/d^2 are then

$$\varepsilon = \frac{[N_1\lambda_o^2 + N_2^{-1}(1 - N_1\lambda_o)^2]^2}{N_1\lambda_o^3 + N_2^{-2}(1 - N_1\lambda_o)^3} \qquad (5.82a)$$

and

$$\frac{\alpha_p}{\alpha} = \frac{N_1\lambda_o^3 + N_2^{-2}(1 - N_1\lambda_o)^3}{N_1\lambda_o^2 + N_2^{-1}(1 - N_1\lambda_o)^2}. \qquad (5.82b)$$

In particular, for sufficiently large values of N_2, we have

$$\frac{\alpha_p}{\alpha} \approx \lambda_o \qquad (5.83a)$$

and

$$\varepsilon \approx N_1\lambda_o. \qquad (5.83b)$$

It is worth noting that the error-probability exponent actually converges to the value determined from this bound when N_2 is sufficiently large. That is, the system in fact performs as an equal-strength diversity system with an energy–to–noise ratio per diversity path of $\alpha\lambda_o$ and with ε $(= N_1\lambda_o)$ times the energy–to–noise ratio of the given system. The validity of this statement can be established by an application of (5.26) and (5.27).

We observe that, although ε equals unity when all the positive eigenvalues are equal to each other, the presence of a very large number of exceedingly small eigenvalues may appreciably diminish the value of ε. Specifically if a fraction $N_1\lambda_o$ of the total energy is divided equally among N_1 paths while a fraction $1 - N_1\lambda_o$ is divided equally among N_2 paths and if N_2 tends to infinity, the system performs as though the fraction $1 - N_1\lambda_o$ is lost. This occurs because the latter fraction is so finely divided that severe square-law suppression occurs in the demodulator. In fact, as the division is made finer and finer (N_2 increases), the paths in question are "captured" by noise, and the system performs as though only the fraction $N_1\lambda_o$ of the energy were employed.

Upper Bound

It has not been possible to obtain an upper bound that is as strong as the corresponding lower bound, but a rather useful result does exist for information rates that are not too small or equivalently for values of β that are not too large. This bound is used in the subsequent discussion of threshold effects and of transmitter peak power requirements. Although those discussions are contained in Chapter 6, the bound is best presented here.

We recall, from Fig. 5.3, that $f_\rho(x)$ is a concave increasing function of x for x less than $\alpha_p{}^\circ$ and is a decreasing function for x greater than $\alpha_p{}^\circ$. Therefore

$$f_\rho(x) \le f_\rho(y) + (x-y)f'_\rho(y) \qquad (5.84)$$

for all x, provided that y is no larger than $\alpha_p{}^\circ$. In this expression $f'_\rho(y)$ denotes the partial derivative of $f_\rho(y)$ with respect to y.

Upon introducing the right member of (5.84) into the expression for $E(\rho)$ (5.26), we obtain

$$E(\rho) \le f_\rho(y) + (\alpha b - y)f'_\rho(y) \qquad (5.85a)$$

provided that

$$y \le \alpha_p{}^\circ, \qquad (5.85b)$$

where, as before,

$$b = \sum_i \lambda_i{}^2. \qquad (5.85c)$$

To obtain the sharpest result, we minimize the right member of (5.85a) with respect to y. This minimum is attained when y equals αb, unless αb exceeds $\alpha_p{}^\circ$, in which case it is attained when y equals $\alpha_p{}^\circ$. Thus

$$\begin{aligned} E(\rho) &\le f_\rho(\alpha b), &\quad \text{if } \alpha b \le \alpha_p{}^\circ, \\ E(_\rho) &\le f_\rho(\alpha_p{}^\circ), &\quad \text{if } \alpha b \ge \alpha_p{}^\circ. \end{aligned} \qquad (5.86)$$

Equation 5.86, in conjunction with (5.76), yields the desired upper bound to the system reliability. We now state these results in terms of E_b, the reliability per information bit; the corresponding statements in terms of E are readily obtained.

Consider a fading dispersive system operating with given values of v, b, β, and α ($=v\beta$). Let $\alpha_p{}^\circ$ be the optimum energy–to–noise ratio per diversity path in an equal-strength system employing the given value of β. If αb exceeds $\alpha_p{}^\circ$, the reliability of the given system is upper bounded by that of the optimized equal-strength diversity system employing the same values of v and β. If αb is less than $\alpha_p{}^\circ$, the reliability of the given system is upper bounded by that of the equal-strength diversity system employing the given values of v and β, and an energy–to–noise ratio per diversity path of αb.

The latter of the two bounds prevails for all values of β if αb is less than 3, since it then is less than $\alpha_p{}^\circ$ for all values of β. If αb exceeds 3, the reliability of the given system is upper bounded by the reliability of an optimized equal-strength system for sufficiently large values of β. However as β is decreased, the value of $\alpha_p{}^\circ$ increases until it equals αb, and any further decrease in β takes the system into the region where E_b is bounded by the E_b of the non-optimized equal-strength system. Therefore for sufficiently small values of

β, the reliability of any given system is no greater than that of an equal-strength diversity system operating at an energy–to–noise ratio per diversity path of αb.

REFERENCES

[1] J. M. Wozencraft and I. M. Jacobs, *Principles of Communication Engineering*. New York: Wiley, 1965, pp. 233–245.

[2] D. J. Sakrison, *Communication Theory: Transmission of Waveforms and Digital Information*. New York: Wiley, 1968, pp. 200 and 240.

[3] C. W. Helstrom, *Statistical Theory of Signal Detection*. New York: Pergamon, 1960, pp. 91–95.

[4] R. G. Gallager, *Information Theory and Reliable Communication*. New York: Wiley, 1968, pp. 379–381.

[5] J. M. Wozencraft and I. M. Jacobs, *Principles of Communication Engineering*. New York: Wiley, 1965, pp. 342–346.

[6] R. M. Fano, *Transmission of Information*. New York: Wiley, and M.I.T. Press, 1961, p. 202.

[7] C. E. Shannon, "Probability of Error for Optimal Codes in a Gaussian Channel." *Bell System Tech. J.*, 611, May 1959.

[8] R. M. Fano, *Transmission of Information*. New York: Wiley, and M.I.T. Press, 1961, pp. 332–334.

[9] R. G. Gallager, "A Simple Derivation of the Coding Theorem and Some Applications." *IEEE Trans. Inform. Theory*, pp. 3–18, January 1965.

[10] J. N. Pierce, "Approximate Error Probabilities for Optimal Diversity Combining." *IEEE Trans. Communic. Systems*, pp. 352–354, September 1963.

[11] H. L. Van Trees, *Detection, Estimation, and Modulation Theory*. New York: Wiley, 1970, pt. 2, Chapters 3, 4.

[12] L. D. Collins, "Closed-Form Expressions for the Fredholm Determinant for State-Variable Covariance Functions." *Proc. IEEE*, 350–351, March 1968.

[13] A. B. Baggeroer, "A Finite-Variable Technique for Solving Fredholm Integral Equations," M.I.T., Research Laboratory of Electronics, Technical Report 459, November 15, 1967.

[14] J. N. Pierce, "Theoretical Limitations on Frequency and Time Diversity for Fading Binary Transmissions." *IRE Trans. on Commun. Systems*, pp. 186–189, June 1961.

[15] I. Jacobs, "The Asymptotic Behavior of Incoherent M-ary Communication Systems." *Proc. IEEE*, pp. 251–252, January 1963.

[16] R. G. Gallager, *Information Theory and Reliable Communication*. New York: Wiley. 1968, Chapter 6–8.

[17] J. M. Wozencraft and I. M. Jacobs, *Principles of Communication Engineering*. New York: Wiley, 1965, Chapter 6.

[18] W. W. Peterson, *Error-Correcting Codes*. New York: Wiley, and M.I.T. Press, 1961

[19] E. R. Berlekamp, *Algebraic Coding Theory*. New York: McGraw-Hill, 1968.

[20] A. Kohlenberg and A. S. Berner, "An Experimental Comparison of Coding vs. Frequency Diversity for HF Telegraphy Transmission." *IEEE Trans. Commun. Tech.*, p. 532, August 1966.

[21] Robert G. Gallager, *Information Theory and Reliable Communication*. New York: Wiley, 1968, pp. 82–91.

Chapter 6

Waveforms and Eigenvalues

The results of Chapter 5 depend upon the channel scattering function and modulator waveforms only through the fractional path strengths or eigenvalues λ_i. We now attempt to relate those results directly to the scattering function and modulation envelope. Our primary objective is to determine the performance levels that can be attained with any given scattering function and to establish the waveforms that attain them. We have not been entirely successful in this endeavour; the extent to which the optimum results of Chapter 5 can be approached with an arbitrary channel is not yet known. However it has been possible to conclude that those levels can often be approached to within at least "a few decibels." It has also been shown that the channel capacity does not depend upon the scattering function.

To arrive at these conclusions it has been necessary to pursue several lines of inquiry. Thus in Section 6.1 we employ the diversity estimates of Chapter 3, whereas in Section 6.2 we show that these estimates are rarely precisely correct. The remainder of the chapter focuses upon special situations pertaining either to the scattering function, for example, slightly dispersive, or to the modulation, for example, highly dispersive in a single dimension. In each situation we obtain results that suggest that the optimum performance levels of Chapter 5 can be approached for both overspread and underspread channels. Moreover all of the examples suggest that the diversity estimates of Section 3.6 provide a reasonable compromise between simplicity and precision in system design; that is, the system suggested by these estimates often performs as well as that resulting from a more sophisticated analysis. However the performance levels themselves cannot be determined adequately from the estimates.

148

6.1 DIVERSITY ESTIMATES

The simplest but least precise estimate of the fractional path strengths is provided by the heuristic arguments of Section 3.6; that is, we suppose that all of the path strengths are equal and estimate the number D of paths from the results of Section 3.6. For example, if the scattering function is not skewed and is concentrated in time and frequency (Fig. 3.9) and if TW equals one, we have

$$D = (1 + BT)(1 + LW). \tag{6.1}$$

Alternately, if we are certain that the path strengths are equal, we know from (3.46) that

$$D = \left[\iint |\mathcal{R}(\alpha, \beta)\theta(\beta, \alpha)|^2 \, d\alpha \, d\beta \right]^{-1}, \tag{6.2}$$

where $\mathcal{R}(\alpha, \beta)$ is the two-frequency correlation function of the channel and $|\theta(\beta, \alpha)|^2$ is the ambiguity function.

To the extent that the approximations of Section 3.6 are valid, the performance of a system with a given modulation envelope and scattering function can be determined from the D-fold equal strength results of Chapter 5. In particular, the search for an "optimum" waveform reduces to the determination of the values of T and W for which D equals D° as computed from (5.42) and (5.45). For example, if β exceeds 12 db, T and W should be chosen so that D equals $\alpha/3$, where α is the energy-to-noise ratio.

If one of the simpler expressions for D is applicable, for example, (6.1), the determination of the required values of T and W is trivial. Of course we would be rather surprised if all the waveforms with the given value of T and W yielded precisely the same error probability. However the use of this estimate as a preliminary to more complex performance estimates may result in a considerable reduction in the total design effort.

It is possible to extend the simple estimate of (6.1) to waveforms or scattering functions that are not concentrated in time and frequency [1, 2]. This extension is particularly simple, and important, when the basic complex modulation $u(t)$ is composed of a sum of constituent parts $u_i(t)$ whose contributions to the received waveform are statistically independent of each other and do not interfere with each other, that is, when *explicit diversity* is employed. These conditions are just those discussed in Section 4.4. As stated there, they are approximately satisfied if the different $u_i(t)$ are separated at least by either B Hz in frequency or L sec in time and also at least by either L^{-1} Hz in frequency or B^{-1} sec in time. As noted in Section 4.4, when the aforestated conditions are satisfied the eigenvalues, or fractional path strengths, of the complete system are simply related to those associated with the constituent parts of the waveform.

Specifically let the basic complex modulation envelope be

$$u(t) = \sum_{i=1}^{n} \sqrt{p_i} \, u_i(t) \tag{6.3a}$$

with

$$p_i \geq 0, \qquad i = 1, \cdots, n, \tag{6.3b}$$

and

$$\sum_{i=1}^{n} p_i = 1. \tag{6.3c}$$

Also let the independence and noninterference conditions of the preceding paragraph be satisfied. The fractional path strengths of a system employing $u(t)$ are $\lambda_i^j p_j, j = 1, \ldots, n, i = 1, \ldots$, where the $\lambda_i^j, i = 1, \ldots$, are the fractional path strengths that result from the use of $u_j(t)$ alone and p_j is the fraction of the total transmitted energy contributed by $\sqrt{p_j} u_j(t)$. Thus the fractional path strengths associated with $u(t)$ can be approximately determined from estimates of the λ_i^j.

For example, suppose that the use $u_j(t)$ as a complex envelope results in about D_j equal-strength paths, as estimated from (6.1). Then the use of $u(t)$ results in about D diversity paths of fractional strength p_j/D_j for $j = 1, \ldots, n$. In particular if

$$p_j = \frac{D_j}{\sum_i D_i}, \tag{6.4}$$

all of the path strengths associated with $u(t)$ equal $\left(\sum_i D_i\right)^{-1}$. This choice of the p_j is a natural one since equality of the fractional path strengths is necessary to achieve the optimum performance levels of Chapter 5.

The conclusion concerning explicit diversity path strengths can also be reached from (6.2). Specifically if the contributions of the various $u_j(t)$ to the received waveform are statistically independent, the complex correlation function associated with $u(t)$ can be expressed as

$$R(t, \tau) = \sum_i p_j R_j(t, \tau),$$

where $R_j(t, \tau)$ is the complex correlation function associated with $u_j(t)$. Moreover if the responses do not interfere,

$$\int |R(t, \tau)|^2 \, dt \, d\tau = \sum_j p_j^2 \int |R_j(t, \tau)|^2 \, dt \, d\tau.$$

But to the extent that we accept (6.2), this expression implies that

$$D = \left(\sum_j p_j^2 D_j^{-1} \right)^{-1}$$

or if the p_j are given by (6.4)

$$D = \sum_j D_j.$$

Thus we again conclude that the fractional path strengths associated with $u(t)$ are equal to $\left(\sum_j D_j \right)^{-1}$ when the p_j are given by (6.4).

Our approach thus far has been to suppose that a system performs as an equal-strength diversity system and then to obtain a gross estimate of the number of diversity paths it contains. However as we show in Section 6.2, it is doubtful that this performance can be precisely realized for realistic scattering functions; thus although the supposition of equal strengths is an expedient one, it is not obvious that it is justifiable.

Faced with this situation we would ideally return to the general results of Chapter 5 and attempt to determine the complex modulation envelope that yields the best performance for a given channel. Unfortunately this determination is exceedingly difficult. Moreover although the performance of an optimized equal-strength diversity system is perhaps not precisely attainable, it can sometimes be approached very closely with realistic waveforms. To the extent that such an approach is possible, the results of Section 5.5 should provide sharp estimates of the system performance. Thus we concentrate on those results. However before proceeding with the development, we establish the conditions that must be satisfied if a given system is to perform precisely as an equal-strength diversity system. Since these conditions are not employed subsequently, the section may be omitted without loss of continuity.

6.2 CONDITIONS FOR EQUAL EIGENVALUES

We observe that

$$b = \sum_i \lambda_i^2 \le \lambda_o \sum_i \lambda_i = \lambda_o, \tag{6.5}$$

where λ_o, called the dominant eigenvalue, denotes the maximum value of the λ_i. Equality occurs in (6.5) if and only if all of the positive λ_i equal λ_o. Consequently a necessary and sufficient condition for equality of the eigenvalues, and hence for a system to perform as an equal-strength diversity system, is

$$\frac{b}{\lambda_o} = 1. \tag{6.6}$$

By virtue of (2.42) or (3.46), b may also be expressed as

$$b = \iint |R(t, \tau)|^2 \, dt \, d\tau, \tag{6.7}$$

where $R(t, \tau)$ is the complex correlation function associated with the channel scattering function and with the basic complex modulation envelope $u(t)$. Equivalently, and more conveniently,

$$b = \iint |\mathcal{R}(x, y)\theta_u(y, x)|^2 \, dx \, dy, \tag{6.8}$$

where $\mathcal{R}(x, y)$ is the two-frequency correlation function of the channel and where $|\theta_u(y, x)|^2$ is the ambiguity function of $u(t)$, that is,

$$\theta_u(y, x) = \int u\left(t - \frac{y}{2}\right) u^*\left(t + \frac{y}{2}\right) \exp j2\pi xt \, dt. \tag{6.9}$$

The subscript u is added to avoid subsequent confusion.

We next relate the dominant eigenvalue λ_0 to the complex correlation function $R(t, \tau)$ by employing the extremal property of the dominant eigenvalue [3–5]. Specifically

$$\lambda_0 = \max \left[\iiint f(t)R(t, \tau)f^*(\tau) \, dt \, d\tau \right], \tag{6.10a}$$

where the maximization is over all unit norm functions $f(t)$, that is, all functions such that

$$\int |f(t)|^2 \, dt = 1. \tag{6.10b}$$

After some manipulation (6.10a) can be restated as

$$\lambda_0 = \max \left[\iint \mathcal{R}(x, y)\theta_u(y, x)\theta_f(y, -x) \, dx \, dy \right], \tag{6.11}$$

where the maximization is again over all unit norm functions $f(t)$ and where $\theta_f(\cdot, \cdot)$ is determined from (6.9) with $f(t)$ substituted for $u(t)$.

The dominant eigenvalue λ_0 equals or exceeds the integral of (6.11) for every function $f(t)$. Also by virtue of (6.5), λ_0 equals or exceeds b. Consequently b/λ_0 equals unity if and only if b equals or exceeds the right member of (6.11) for every unit norm function $f(t)$. Therefore a necessary and sufficient condition for equality of the positive eigenvalues is

$$\iint |\mathcal{R}(x, y)\theta_u(y, x)|^2 \, dx \, dy \geq \iint \mathcal{R}(x, y)\theta_u(y, x)\theta_f(y, -x) \, dx \, dy \tag{6.12}$$

for every unit norm function $f(t)$.

The condition of (6.12) is particularly useful in determining classes of channels for which equality of the positive eigenvalues cannot be achieved. For example, a necessary condition is that (6.12) be satisfied with $f(t)$ set equal to $u^*(t)$. This choice of $f(t)$ yields the necessary condition

$$\iint [|\mathscr{R}(x, y)|^2 - \mathscr{R}(x, y)]|\theta_u(y, x)|^2 \, dx \, dy \geq 0. \tag{6.13}$$

A surprising number of two-frequency correlation functions fail to satisfy (6.13). Many, if not all, real-valued nonnegative two-frequency correlation functions fail to satisfy it. In fact, if $\mathscr{R}(x, y)$ is real-valued and nonnegative,

$$|\mathscr{R}(x, y)|^2 - \mathscr{R}(x, y) \leq 0 \qquad \text{for all } x \text{ and } y, \tag{6.14}$$

and (6.13) cannot be satisfied unless the integrand vanishes identically. This in turn requires that

$$\theta_u(y, x) = 0 \tag{6.15a}$$

for all x and y such that

$$\mathscr{R}(x, y) \neq 1. \tag{6.15b}$$

The condition of (6.15) is a very stringent one. Indeed both $\mathscr{R}(x, y)$ and $\theta_u(y, x)$ equal unity at the origin, and $\theta_u(y, x)$ is a continuous function of x and y if $u(t)$ is bounded. Consequently the condition is violated unless $\mathscr{R}(x, y)$ is identically unity in "the vicinity" of the origin. It is doubtful that such a two-frequency correlation function exists.

6.3 PRELIMINARIES

Our objectives are to bound the performance levels that can be attained with any given channel and to establish the character of the modulation waveforms that realize these levels. For the greater part of the discussion the performance is estimated by the bounds of (5.17), (5.77), and (5.78) That is

$$P(\varepsilon) \leq K_2 \cdot 2^{-\nu E_b}, \tag{6.16}$$

where

$$E_b = \max_{0 \leq \rho \leq 1} \left[\frac{\varepsilon\beta}{\ln 2} f_\rho(\alpha_p) - \rho \right], \tag{6.17}$$

$$\varepsilon = \sqrt{Db}, \tag{6.18a}$$

$$\alpha_p = \alpha \left(\frac{b}{D} \right)^{1/2} = \frac{\varepsilon\alpha}{D} \tag{6.18b}$$

and where D may be chosen as any number that satisfies the inequality

$$D \leq \frac{b^3}{d^2} \qquad (6.19a)$$

with

$$b = \sum_i \lambda_i^2 \qquad (6.19b)$$

and

$$d = \sum_i \lambda_i^3. \qquad (6.19c)$$

As in Chapter 5, we focus on the exponent E_b to the exclusion of the coefficient K_2. Although this exponent is maximized by setting D to equal b^3/d^2, our primary objective is to obtain simple results and that objective is sometimes best served by employing smaller values for D.

We recall that v, α and β are, respectively, the number of information bits conveyed per transmission, the average energy-to-noise ratio per transmission, and the average energy-to-noise ratio per information bit. We also recall that a system with given values of v, α, β, b, and d performs at least as well (exponentially) as any equal-strength diversity system with the given value of v, an energy-to-noise ratio per diversity path of α_p, and an energy-to-noise ratio per information bit of $\varepsilon\beta$.

As in Chapter 5 we call D the effective diversity, α_p the effective-energy–to–noise ratio per diversity path and $\varepsilon\beta$ the effective-energy–to–noise ratio per information bit. Of course these are really only bounds to the effective quantities, both because they depend upon our choice of D and because a given system may perform better than an equal-strength system with an energy-to-noise ratio of $b^2\beta/d$ per information bit. Although this weakness in our result is certainly undesirable, it appears to be a necessary compromise if we are to obtain simple analytical results. Moreover to the extent that we find waveforms for which ε is approximately equal to one, we are assured that our results are sharp. The only alternative now available is numerical analysis. Very efficient computational algorithms are available for some restricted classes of problems, for example, singly spread channels [6].

We evaluate both D and ε, but the greater emphasis is placed on the quantity ε because D can be adjusted independently of ε and hence can be optimized; that is, no matter what the values of v, β, α, b, and d, we can alter the modulation so that E_b is at least as large as it is for an *optimized* equal-strength diversity system with an energy-to-noise ratio per information bit of $\varepsilon\beta$. The opportunity to control the effective diversity is afforded by the use of explicit diversity as described in Sections 4.4 and 6.1. This technique enables us to approach the performance levels of Chapter 5 for a much broader class of channels than is otherwise possible.

In particular, explicit diversity provides a means of splitting eigenvalues, or effective diversity paths. Precisely stated, given a set of eigenvalues λ_1, $\lambda_2, \ldots$ associated with a scattering function $\sigma(r, f)$ and modulator waveform $u(t)$, we can construct a modulator waveform that causes each λ_i to be replaced by a set of D_e eigenvalues, each of value $\lambda_i D_e^{-1}$; that is, each λ_i is replaced by an eigenvalue of multiplicity D_e and of value $\lambda_i D_e^{-1}$. This can be accomplished by forming D_e time or frequency translates of $u(t)$ with the translations chosen to satisfy the independence and noninterference conditions discussed in Sections 4.4 and 6.1. The sum of these waveforms, multiplied by $D_e^{-1/2}$ (to retain the unit normalization), is a waveform that yields the desired set of eigenvalues.

Explicit diversity provides a means of varying the effective diversity without affecting the value of ε. Specifically if b and d are the quantities of (6.19b) and (6.19c) for the original system, the corresponding values after splitting are bD_e^{-1} and dD_e^{-2}. Thus the value of ε for the resulting system is

$$\varepsilon = \left(\frac{Db}{D_e}\right)^{1/2}, \tag{6.20a}$$

where the choice of D, the "effective" diversity of the resulting system, is subject only to the constraint

$$D \le \frac{b^3}{d^2} D_e. \tag{6.20b}$$

A more compact result is obtained by defining

$$D_i = \frac{D}{D_e} \tag{6.21a}$$

which yields

$$\varepsilon = \sqrt{D_i b}, \tag{6.21b}$$

where D_i is subject to the constraint

$$D_i \le \frac{b^3}{d^2} \tag{6.21c}$$

and where b and d are given by (6.19b) and (6.19c).

Equation 6.21a presents the total effective diversity D as the product of two factors, D_e and D_i. We refer to D_i as the diversity implicit in the original system and to D_e as the explicit diversity. The significance of explicit diversity and the means of its realization suggest that the resulting systems be called D_e-fold explicit diversity systems based upon the modulation $u(t)$. This nomenclature is consistent with the usual concept of a diversity system.

It is important to note that the value of ε depends upon our choice of D_i. Since we subsequently have occasion to use several different values of D_i, we should explicitly indicate the dependence of ε upon this choice. However this would result in a cumbersome notation, and so we choose instead to use the unmodified symbol ε with the understanding that the value of D_i is specified by the context.

Equations 6.21 clearly indicate that we can vary the total effective diversity $D_e D_i$ without altering the value of ε by varying the explicit diversity D_e. In particular we may optimize (6.17) with respect to α_p by choosing D_e so that α_p equals $\alpha_p{}^\circ$, that is,

$$D_e D_i = \frac{\varepsilon\alpha}{\alpha_p{}^\circ},\tag{6.22}$$

where $\alpha_p{}^\circ$ is the optimum energy-to-noise ratio per diversity path for the given value of $\varepsilon\beta$. The derived system then performs at least as well as an optimized equal-strength diversity system with an energy-to-noise ratio per information bit of $\sqrt{D_i b}$ times that of the given system. That is, $\sqrt{D_i b}$ measures the efficiency, relative to an optimized equal-strength system, of a system employing a given channel and basic modulator waveform with an optimum amount of explicit diversity. There are, of course, threshold conditions that must be satisfied before the efficiency $\sqrt{D_i b}$ can be realized. For example, $\varepsilon\alpha/D_i$ must exceed $\alpha_p{}^\circ$, otherwise the best performance is realized when no explicit diversity is employed. These problems are analogous to those encountered in Chapter 5 and require no further comment here.

In principle our next objective should be to determine the maximum value of ε, as computed from (6.21), for any given channel. Although this maximization can be performed numerically for any specific channel, it is far too complicated to yield results of general utility. Therefore we instead consider several limiting situations that provide some understanding of the performance levels that can be attained. Specifically we estimate the efficiency that results when we employ waveforms for which the given channel appears to be either only slightly dispersive, highly singly dispersive, or highly doubly dispersive.

6.4 SLIGHTLY DISPERSIVE CHANNELS

We now evaluate the bounds of (6.21) under the supposition that the channel is only slightly dispersive. Specifically we first determine the form of (6.21) when the channel is nondispersive; we then choose the value of D_i so as to simplify the expression for ε without weakening it appreciably for channels that appear to be only slightly dispersive. The character of the result is illustrated by two examples.

In Section 3.3 we define a channel to be nondispersive if its scattering function is an impulse in time and frequency, that is, if

$$\sigma(r, f) = \delta(r - r_o)\delta(f - f_d), \tag{6.23}$$

where $\delta(\cdot)$ denotes the unit-impulse function. The complex correlation function of such a channel is

$$R(t, \tau) = u(t - r_o)u^*(\tau - r_o) \exp j2\pi f_d(\tau - t). \tag{6.24}$$

It is easy to verify that $u(t - r_o) \exp -j2\pi f_d t$ is an eigenfunction of this correlation function and that the associated eigenvalue is unity. By virtue of Mercer's theorem (2.42), this implies that $R(t, \tau)$ possesses the single positive eigenvalue of unity; hence any nondispersive channel behaves as a single path diversity system—a conclusion that is hardly surprising.

When the dominant eigenvalue λ_o of a system is only slightly less than unity, we say that the system is only *slightly dispersive*. All but the dominant eigenvalue are then much less than unity and ε, as given by (6.21) with D_i set equal to b^3/d^2, becomes approximately

$$\varepsilon \approx \frac{b}{\lambda_o} \approx \sqrt{b}, \tag{6.25}$$

where we have used the approximations

$$d \approx \lambda_o b, \tag{6.26a}$$

$$\lambda_o \approx \sqrt{b}. \tag{6.26b}$$

The rightmost member of (6.25) provides a valid expression for ε even when the approximations of (6.26) are not satisfied. To see this, we note that $d \leq \lambda_o b$ and $b \geq \lambda_o^2$ and so

$$\frac{b^3}{d^2} \geq \frac{b}{\lambda_o^2} \geq 1. \tag{6.27}$$

Thus (6.21c) is satisfied if

$$D_i = 1 \tag{6.28a}$$

which yields

$$\varepsilon = \sqrt{b}. \tag{6.28b}$$

Of course the attainment of this efficiency requires that an appropriate amount of explicit diversity be employed; otherwise the efficiency may be far less than $\sqrt{b}$.

It is instructive to determine the value of b, and hence of ε, for a few specific systems. As a first example let us suppose that a chirped Gaussian-shaped pulse is employed in the modulator, that is,

$$u(t) = \frac{2^{1/4}}{\sqrt{T}} \exp - \pi\left(\frac{t}{T}\right)^2 [1 + j\sqrt{(TW)^2 - 1}], \qquad (6.29a)$$

where T and W are the duration and bandwidth of $u(t)$ as defined by (3.2). Let us further suppose that the channel scattering function has the shape of a two-dimensional Gaussian probability density; that is

$$\sigma(r, f) = \frac{2}{S} \exp - \frac{2\pi}{S^2} [(rB)^2 + (fL)^2 - 2rf\sqrt{(BL)^2 - S^2}], \qquad (6.29b)$$

where B, L, and S are, respectively, the Doppler spread, multipath spread, and total spread of the channel as defined by (3.1).

These expressions for $u(t)$ and $\sigma(r, f)$, in conjunction with (2.22) and (2.25), yield the following equations for $\theta(\tau, f)$ and for the two-frequency correlation function $\mathscr{R}(\alpha, \beta)$:

$$\theta(\tau, f) = \exp - \frac{\pi}{2}\left[(\tau W)^2 + (fT)^2 + 2f\tau\sqrt{(TW)^2 - 1}\right], \qquad (6.30a)$$

$$\mathscr{R}(\alpha, \beta) = \exp - \frac{\pi}{2}\left[(\alpha L)^2 + (\beta B)^2 + 2\alpha\beta\sqrt{(BL)^2 - S^2}\right]. \qquad (6.30b)$$

Upon introducing these expressions into (6.8) and evaluating the resulting integral, we obtain

$$b = \left[1 + S^2 + (BT)^2 + (LW)^2 - 2\sqrt{(TW)^2 - 1}\sqrt{(BL)^2 - S^2}\right]^{-1/2}. \qquad (6.31a)$$

Hence, by virtue of (6.28b),

$$\varepsilon = \left[1 + S^2 + (BT)^2 + (LW)^2 - 2\sqrt{(TW)^2 - 1}\sqrt{(BL)^2 - S^2}\right]^{-1/4}. \qquad (6.31b)$$

Let us now suppose that we are free to vary T and W, and let us choose them so as to maximize ε. It is easy to show that for any fixed value, say $\hat{K}$, of the product TW, the maximum is attained when

$$T = \sqrt{\frac{L\hat{K}}{B}}, \qquad (6.32a)$$

and the resulting maximized value of ε is

$$\varepsilon = \left[1 + S^2 + 2BL\hat{K} - 2\sqrt{\hat{K}^2 - 1}\sqrt{(BL)^2 - S^2}\right]^{-1/4}. \qquad (6.32b)$$

Upon maximizing this expression with respect to $\hat{K}$, we find that $\hat{K}$ should be set equal to BL/S. Thus ε is maximized when

$$T = \frac{L}{\sqrt{S}}, \tag{6.33a}$$

$$W = \frac{B}{\sqrt{S}}, \tag{6.33b}$$

and the resulting optimized value of ε is

$$\varepsilon = \frac{1}{\sqrt{1+S}}. \tag{6.33c}$$

As a second example, let us suppose that

$$u(t) = \left(\frac{2}{T}\right)^{1/2} \exp - \frac{t}{T}, \qquad t \geq 0,$$

$$u(t) = 0, \qquad t < 0, \tag{6.34a}$$

and that

$$\sigma(r,f) = \frac{4}{BL} \frac{\exp - 2(r/L)}{1 + (2\pi f/B)^2}, \qquad r \geq 0,$$

$$\sigma(r,f) = 0, \qquad r < 0. \tag{6.34b}$$

For this scattering function, S equals BL.

The expressions for $\theta(\tau, f)$ and $\mathscr{R}(\alpha, \beta)$ are readily evaluated for the system described by (6.34a) and (6.34b). The introduction of these expressions into (6.8) yields

$$b = \left(1 + BT + \frac{L}{T} + BL\right)^{-1}. \tag{6.35}$$

The value of b is maximized by setting T equal to $\sqrt{L/B}$ or, since S equals BL, by setting

$$T = \frac{L}{\sqrt{S}}. \tag{6.36a}$$

The resulting value of b is $\left(1 + \sqrt{S}\right)^{-2}$; hence

$$\varepsilon = \frac{1}{1 + \sqrt{S}}. \tag{6.36b}$$

There are many more examples for which b can be evaluated in closed form. In many of these the resulting value of b is approximately maximized when T

is set equal to $L/\sqrt{S}$, and the resulting bound to ε is often approximately equal to the value computed from (6.33c) or (6.36b). Thus it appears that ε can be made relatively large if S does not exceed unity and that, lacking other information, it may be reasonable to set T equal to $L/\sqrt{S}$. As a rough rule of thumb, we may say that ε exceeds 0.5 if S does not exceed unity. Stated alternately, the loss in effective energy is at most about 3 db if S does not exceed unity.

It is worthwhile to consider the weakening in the result that may have occured in the transition from (6.21) to (6.28). To this end, the expression of (6.21b) is evaluated for the first example considered above. This evaluation is possible because the eigenvalues of the resulting complex correlation function can be determined.

For the Gaussian-shaped $u(t)$ and the Gaussian-shaped scattering function of (6.29), it follows from (2.16) that

$$R(t, \tau) = \sqrt{2\mu_2}\, \exp -\frac{\pi}{2}[(\tau - t)^2(\mu_1 - \mu_2\mu_3^2)$$

$$+ (\tau + t)^2\mu_2 + 2j(t^2 - \tau^2)\mu_2\mu_3], \qquad (6.37)$$

where

$$\mu_1 = B^2 + W^2, \qquad (6.38a)$$

$$\mu_2 = (L^2 + T^2)^{-1}, \qquad (6.38b)$$

and

$$\mu_3 = \sqrt{(TW)^2 - 1} + \sqrt{(BL)^2 - S^2}. \qquad (6.38c)$$

This correlation function is known to possess the eigenvalues [7]

$$\lambda_i = (1 - c)c^{i-1}, \qquad i = 1, \ldots, \qquad (6.39a)$$

where c is uniquely determined from the expression

$$\left(\frac{1+c}{1-c}\right)^2 = 1 + (BT)^2 + (LW)^2 + S^2 - 2\sqrt{(TW)^2 - 1}\sqrt{(BL)^2 - S^2}.$$

$$(6.39b)$$

Consequently the summations for b and d reduce to geometric series that can be evaluated to obtain

$$b = \frac{1-c}{1+c} \qquad (6.40a)$$

and

$$d = \frac{(1-c)^3}{1-c^3}. \qquad (6.40b)$$

Upon setting D_i equal to b^3/d^2 and introducing (6.40) into (6.21b) we obtain the following inequality:

$$\varepsilon = \frac{1 + c + c^2}{(1 + c)^2} \tag{6.41}$$

We next maximize the right member of (6.41) with respect to the waveform duration T and bandwidth W. By straightforward calculation we find that the maximum is obtained when c is minimized, that is, when

$$T = \frac{L}{\sqrt{S}} \tag{6.42a}$$

and

$$W = \frac{B}{\sqrt{S}}. \tag{6.42b}$$

The resulting bound to ε is then

$$\varepsilon = \frac{3}{4} + \frac{1}{4}\left(\frac{1}{1 + S}\right)^2. \tag{6.43}$$

It is worth noting that the values of T and W that maximize ε, as given by (6.41), also maximize $\sqrt{b}$. Moreover for values of S less than unity, the bounds of (6.33c) and (6.43) do not differ appreciably; for example, for a unit spread factor, (6.33c) yields a value of 0.707 whereas (6.43) yields 0.81. However as S increases, the bound of (6.33c) vanishes whereas that of (6.43) approaches a limiting value of 0.75. Thus the result of (6.33c) is very weak for channels that have large spread factors.

6.5 HIGHLY SINGLY DISPERSIVE SYSTEMS

In the previous discussion we sought estimates of the performance that results when an attempt is made to employ a channel so that it "appears" to be nondispersive. In the following discussion we consider the performance that results when an attempt is made to have a channel "appear" to be highly dispersive in a single dimension. A highly dispersive mode of operation can be advantageous for either of two reasons. First, for a given channel, it may permit us to increase the performance levels over those that result when we attempt to minimize the amount of implicit diversity. Second, it may significantly reduce the amount of explicit diversity required to achieve a given level of performance, thereby simplifying the implementation of the system.

We first seek to make a channel appear to be nondispersive in time but highly dispersive in frequency. The results of that analysis are then extended to other highly singly dispersive situations.

Highly Frequency Dispersive Systems

According to the heuristic discussion of Section 3.3, a system appears to be dispersive only in frequency if LW is much less than unity. If BT is much greater than unity for such a system, we say that it is highly frequency dispersive. A system involving any given scattering function can always be made highly frequency dispersive by employing a modulation envelope with a sufficiently large duration T and a sufficiently small bandwidth W.

The expressions for ε and D_i are relatively simple for highly frequency dispersive systems; at least they become simple as T increases without limit while the TW product remains constant. We present here the asymptotic expressions for ε and for the true error-probability exponent in the limit of increasing T. We also determine some expressions for ε that are valid for all values of T.

For most of the discussion we suppose that $u(t)$, the complex modulation envelope, is of the form

$$u(t) = \frac{1}{\sqrt{T}} u_o\left(\frac{t}{T}\right), \tag{6.44a}$$

where $u_o(t)$ is an arbitrary waveform of unit norm and unit time duration, that is,

$$\int |u_o(t)|^2 \, dt = \int |u_o(t)|^4 \, dt = 1. \tag{6.44b}$$

The duration of $u(t)$ is then T and its bandwidth is W_o/T, where W_o is the bandwidth of $u_o(t)$. Our asymptotic expressions are obtained by supposing that T tends to infinity.

Asymptotic Expressions for ε: We seek the limiting expressions for ε and D_i as computed from (6.21), with $D_i = b^3/d^2$, when $u(t)$ is given by (6.44a) and T tends to infinity. These expressions follow from the asymptotic expressions for b and d which are derived in Appendix 4. Specifically

$$\lim_{T \to \infty} bT = \left(\int |u_o(t)|^4 \, dt\right)\left(\int \sigma^2(f) \, df\right) \tag{6.45a}$$

and

$$\lim_{T \to \infty} dT^2 = \left(\int |u_o(t)|^6 \, dt\right)\left(\int \sigma^3(f) \, df\right), \tag{6.45b}$$

where $\sigma(f)$ is the Doppler scattering function, that is,

$$\sigma(f) = \int \sigma(r, f)\, dr. \tag{6.45c}$$

A combination of (6.21), (6.44*b*), and (6.45) yields the following limiting expressions for ε and D_i when D_i is set equal to b^3/d^2:

$$\lim_{T \to \infty} \varepsilon = \varepsilon_\infty \tag{6.46a}$$

and

$$\lim_{T \to \infty} \frac{D_i}{T} = B(\varepsilon_\infty)^2, \tag{6.46b}$$

where

$$\varepsilon_\infty = \frac{[\int \sigma^2(f)\, df]^2}{(\int |u_o(t)|^6\, dt)(\int \sigma^3(f)\, df)} \tag{6.46c}$$

and where B is the channel Doppler spread, that is,

$$B = \left\{ \int [\sigma(f)]^2\, df \right\}^{-1}. \tag{6.46d}$$

Equations 6.46 establish an efficiency that assuredly can be achieved if the complex modulation envelope is given by (6.44*a*) and T is sufficiently large. However this assurance is predicated on the stipulation that $\alpha \varepsilon_\infty / D_e D_i$, the energy-to-noise ratio per effective diversity path, is maintained at the optimum value, α_p^o, for example, 3 for sufficiently low rates. Since D_i is asymptotically proportional to T, this stipulation can be satisfied only if the total-energy–to–noise ratio α is also proportional to T. In particular, if no explicit diversity is employed, the asymptotic form of α must be

$$\alpha = \alpha_p^o T B \varepsilon_\infty. \tag{6.47}$$

One means of satisfying (6.47) is to let the signaling interval τ increase linearly with T. For a fixed information rate R, this implies that the alphabet size m is increasing exponentially with T since m equals $2^{R\tau}$. Fortunately the asymptotic values of ε and D_i are sometimes essentially achieved before T, and hence m, becomes prohibitively large. We return to this question subsequently.

Since the value of ε_∞ is often close to unity, it is frequently convenient to estimate the asymptotic implicit diversity by the expression BT rather than by $BT\varepsilon_\infty^2$. The use of the estimate BT is also suggested by the diversity estimates of Section 6.1, for, under the asymptotic conditions considered here, (6.1) and (6.2) behave asymptotically as BT.

We next maximize ε_∞ with respect to $u_o(t)$. This maximization is equivalent to minimizing the quantity

$$Q = \int |u_o(t)|^6 \, dt \tag{6.48}$$

subject to the constraints of (6.44b). We now show that the minimum value of Q is unity and that the minimum is attained when

$$u_o(t) = \begin{cases} 1, & |t| \le \tfrac{1}{2}, \\ 0, & \text{elsewhere.} \end{cases} \tag{6.49}$$

We note that

$$1 = \left[\int |u_o(t)|^4 \, dt \right] = \left(\int |u_o(t)| \cdot |u_o(t)|^3 \, dt \right)^2, \tag{6.50a}$$

or by virtue of Schwartz's inequality [8],

$$\left[\int |u_o(t)|^4 \, dt \right]^2 \le \left[\int |u_o(t)|^2 \, dt \right] \left[\int |u_o(t)|^6 \, dt \right], \tag{6.50b}$$

where equality prevails if and only if

$$|u_o(t)| = K |u_o(t)|^3 \tag{6.50c}$$

for some constant K. These equations imply that Q equals or exceeds unity. Moreover, Q equals unity whenever (6.50c) is satisfied, that is, whenever the magnitude of $u_o(t)$ is equal to unity over a total, possibly disjoint, time interval of unit length and is equal to zero elsewhere. Thus the minimum value of Q is one. Equation 6.49 defines one of the waveforms for which Q equals one. The corresponding expression for $u(t)$ is

$$u(t) = \begin{cases} \dfrac{1}{\sqrt{T}}, & \text{for } |t| \le \dfrac{T}{2}, \\ 0, & \text{elsewhere.} \end{cases} \tag{6.51}$$

Upon introducing the waveform of (6.49) into (6.46c) we obtain

$$\varepsilon_\infty = \frac{\left[\int \sigma^2(f) \, df \right]^2}{\int \sigma^3(f) \, df}. \tag{6.52}$$

This is the efficiency that we can certainly achieve relative to an optimized equal-strength diversity system by employing the waveform of (6.51) with an exceedingly large value of T.

It is interesting to determine the degree to which nonoptimum waveforms can approach the performance of an optimum one. To gain some insight into this question, we consider a complex modulation envelope of the form

$$u(t) = \frac{2^{1/4}}{\sqrt{T}} \exp - \pi \left(\frac{t}{T}\right)^2. \tag{6.53a}$$

The corresponding expression for $u_0(t)$ is then

$$u_0(t) = (2)^{1/4} \exp - \pi(t)^2, \tag{6.53b}$$

and the expression for ε_∞ becomes

$$\varepsilon_\infty = \sqrt{0.75} \frac{\{\int [\sigma(f)]^2 \, df\}^2}{\int [\sigma(f)]^3 \, df}. \tag{6.53c}$$

Thus the value of ε_∞ for the modulation of (6.53a) is approximately 86 percent of the maximum value. Equivalently we may say that an 0.6 db loss in effective power is incurred by using the modulation of (6.53a) rather than that of (6.51). It is satisfying to note that our results are not unduly sensitive to the modulation employed. This insensitivity has also been noted in another context [9].

The maximum values of the asymptotic efficiency and of the associated implicit diversity are presented in Table 6.1 for a variety of scattering functions. These quantities are evaluated from (6.52) and (6.46b) with the optimum modulation specified by (6.51). Since the results depend only upon the Doppler scattering function, it is the only attribute of the scattering function that is included in the table.

As shown in the first entry of the table, the efficiency is unity for a rectangular Doppler scattering function. Of course, the same efficiency results for any Doppler scattering function that assumes a single value other than zero. These conclusions are important since they prove that "overspread" channels are not inherently inferior to underspread channels. For example, the channel whose scattering function is given by the expression

$$\sigma(r, f) = \begin{cases} \dfrac{1}{BL}, & |r| < \dfrac{L}{2}, |f| < \dfrac{B}{2}, \\[2mm] 0, & \text{otherwise,} \end{cases} \tag{6.54a}$$

possesses the Doppler scattering function

$$\sigma(f) = \begin{cases} \dfrac{1}{B}, & |f| < \dfrac{B}{2}, \\[2mm] 0, & \text{otherwise.} \end{cases} \tag{6.54b}$$

Thus regardless of the spread BL of this channel, it can be made to perform as an optimized equal-strength system by employing an exceedingly long rectangular pulse for the modulating waveform.

Table 6.1 Typical Values of ε_∞ and $\lim_{T\to\infty} (Di/T)$ for Highly Frequency Dispersive Channels

	$\sigma(f)$	$\lim\limits_{T\to\infty} Di/T$	ε_∞
	$B^{-1}, \ \|f\| < \dfrac{B}{2}$	B	1
	$\dfrac{2}{B}\exp-4\dfrac{\|f\|}{B}$	$0.56B$	0.75
	$\dfrac{2}{B}\exp-2\dfrac{f}{B}, f\geq 0$	$0.56B$	0.75
	$\dfrac{\sqrt{2}}{B}\exp-2\pi\left(\dfrac{f}{B}\right)^2$	$0.75B$	0.866
	$\dfrac{\sqrt{2}}{B}\exp-\dfrac{\pi}{2}\left(\dfrac{f}{B}\right)^2, f\geq 0$	$0.75B$	0.866
	$\dfrac{2}{B}\left[1+\left(\dfrac{2\pi f}{B}\right)^2\right]^{-1}$	$0.44B$	0.66
	$\dfrac{1.5}{B}\left(1-\dfrac{3}{4}\dfrac{f}{B}\right), 0\leq\dfrac{f}{B}\leq\dfrac{4}{3}$	$0.77B$	0.88
	$\dfrac{1.5}{B}\left(1-\dfrac{3}{2}\dfrac{\|f\|}{B}\right), \left\|\dfrac{f}{B}\right\|<\dfrac{2}{3}$	$0.77B$	0.88
	$\dfrac{c-1}{(1+f)^c}, f\geq 0$	$\dfrac{(3c-1)^2}{(2c-1)^3}$	$\dfrac{(3c-1)(c-1)}{(2c-1)^2}$

Table 6.1 suggests that ε_∞ can be made relatively large for a variety of scattering functions. However there is no positive lower bound to the minimum value of ε_∞, that is, scattering functions exist for which ε_∞ is arbitrarily small. The last entry describes such a scattering function.

So far nothing has been said about the rate at which ε_∞ is approached. This rate is important because the alphabet size and also the energy-to-noise ratio increase with T. Thus the faster the rate of approach, the smaller the value of T and the more feasible the task of implementing the system.

We can gain some appreciation of the rate of approach by reconsidering the example of (6.37) through (6.41), which involves a Gaussian-shaped pulse used in conjunction with a Gaussian-shaped scattering function. For simplicity we suppose that W equals T^{-1} and that the total spread S equals BL. Equations 6.39b and 6.41 then yield, after some manipulation,

$$\varepsilon = \frac{3}{4} + \frac{1}{4[1 + (BT)^{2} + (L/T)^{2} + (BL)^{2}]}. \tag{6.55}$$

We already know from (6.42) that the right member of (6.55) attains its maximum value when T equals $\sqrt{L/B}$. However if we choose instead to operate in the highly frequency dispersive mode, we see from (6.55) that the asymptotic value of 0.75 is approached roughly as $1/(2BT)^{2}$. Thus BT need not be too large (say 2) before the asymptotic condition is essentially achieved. A comparable conclusion has been reached for those problems to which numerical techniques have been applied [6].

Asymptotic Reliability: We now show how the asymptotic value of the system reliability can be computed without first determining the eigenvalues of the complex correlation function. This result is much stronger than those just discussed, but its strength is usually attainable only with considerable computational effort.

We first recall from Chapter 5 that

$$P(\varepsilon) \approx 2^{-\nu E_b}, \tag{6.56a}$$

where

$$E_b = \max_{0 \le \rho \le 1} \left[\frac{\beta E(\rho)}{\ln 2} - \rho \right], \tag{6.56b}$$

$$E(\rho) = \sum_i \lambda_i f(\alpha \lambda_i), \tag{6.56c}$$

$$f(x) = \frac{1}{x} G(x), \tag{6.56d}$$

and

$$G(x) = \left[(1 + \rho) \ln \left(1 + \frac{\rho x}{1 + \rho} \right) - \rho \ln (1 + x) \right]. \tag{6.56e}$$

We now show that

$$\lim_{T \to \infty} E(\rho) = \frac{1}{\mu} \iint G[\mu |u_o(t)|^2 \sigma(f)] \, dt \, df \tag{6.57a}$$

provided that

$$\alpha = \mu T, \tag{6.57b}$$

where μ is a constant of proportionality that can be adjusted by the use of explicit diversity.

Since the detailed proof of (6.57) is somewhat lengthy, although straightforward, we merely outline the principal steps. Asymptotic expressions for the sums of the squares and cubes of the λ_i have already been introduced [(6.45a) and (6.45b)]. More generally, it is shown in Appendix 4 that

$$\lim_{T \to \infty} \left[\frac{1}{T} \sum_i (T\lambda_i)^k \right] = \left[\int |u_o(t)|^{2k} \, dt \right] \left\{ \int [\sigma(f)]^k \, df \right\} \tag{6.58}$$

for any integer k.

Equation 6.58 leads to a simple asymptote for any expression of the form

$$Q = \frac{1}{T} \sum_i F(\mu T \lambda_i) \tag{6.59a}$$

when

$$F(x) = \sum_{k=1}^{N} a_k x^k, \tag{6.59b}$$

for then

$$Q = \frac{1}{T} \sum_i \sum_{k=1}^{N} a_k (\mu T \lambda_i)^k \tag{6.60a}$$

and by virtue of (6.58),

$$\lim_{T \to \infty} Q = \sum_{k=1}^{N} a_k \left[\int |u_o(t)|^{2k} \, dt \right] \left\{ \int [\sigma(f)]^k \, df \right\} \mu^k$$

or

$$\lim_{T \to \infty} Q = \int\!\!\int F[\mu|u_o(t)|^2 \sigma(f)] \, dt \, df. \tag{6.60b}$$

To obtain (6.57a), we extend (6.60b) to the function $xf(x)$ of (6.56). Since $f(x)$ is a continuous function, the extension can be accomplished by approximating $f(x)$ by a polynomial. Specifically the value of $E(\rho)$, as defined by (6.56c), can be computed to any desired degree of precision by substituting some polynomial for $f(x)$ [10]. The asymptotic value of the expression involving this polynomial may be computed from (6.60b) and the resulting polynomial, with $\mu |u_o(t)|^2 \sigma(f)$ as its argument, may again be approximated by the function $f(\cdot)$. The result is (6.57a).

The asymptotic result of (6.57a) is a very powerful one, but its utility is limited by the difficulty encountered in its application. This difficulty may arise either in the evaluation of (6.57a) or in its subsequent optimization with respect to ρ.

The former difficulty is not particularly severe for any of the scattering functions shown in Table 6.1 provided that they are employed with a rectangular modulating waveform. The latter difficulty seems to occur regardless of the scattering function and modulation. This is not too surprising since the optimization with respect to ρ is rather difficult even when, as in Chapter 5, the eigenvalues are assumed to be equal. Of course, when β is sufficiently large, the optimum value of ρ is unity, and the problem simplifies appreciably. In the following paragraphs we consider several examples with ρ equal to unity.

We recall from Chapter 5 that, when β is sufficiently large,

$$E_b = \frac{\beta E(1)}{\ln 2} - 1. \tag{6.61}$$

In particular, for an optimized equal-strength diversity system

$$E_b^{\,\circ} = 0.216\beta - 1 \tag{6.62}$$

for $\beta \geq 12$ db. To compare the optimum and suboptimum systems, we cast (6.61) in the form

$$E_b = 0.216\beta\varepsilon_t - 1, \tag{6.63a}$$

where

$$\varepsilon_t = \frac{E(1)}{0.216 \ln 2} = \frac{E(1)}{0.15}. \tag{6.63b}$$

It is clear that, for large values of β, the given suboptimum system performs as an optimized equal-strength diversity system employing a β of ε_t times that employed in the given system. Hence it is again appropriate to regard ε_t as the efficiency of the given system relative to the optimized equal-strength system; alternately $\beta\varepsilon_t$ may be interpreted as the effective energy-to-noise ratio per information bit of the given system. Unlike ε, ε_t is an exact, or true, expression; the subscript t is added to emphasize this difference.

The quantity ε_t has been evaluated for several different scattering functions used in conjunction with the rectangular pulse modulation of (6.51). The resulting expressions depend upon the quantity μ/B, which may be interpreted as the energy-to-noise ratio per effective diversity path. Although we can maximize ε_t with respect to μ/B, we choose instead to set μ/B equal to 3—the optimum value for equal eigenvalues. This value may be realized through the use of explicit diversity. The resulting values of ε_t are presented in Table 6.2. The asymptotic lower bound, ε_∞ is also presented for purposes of comparison. Other combinations of channels and scattering functions can also be evaluated and the optimization with respect to μ/B completed [11, 12].

Table 6.2 Asymptotic Efficiencies for Highly Frequency Dispersive Channels

$\sigma(f)$	ε_t	ε_∞
$\dfrac{1.5}{B}\left(1 - \dfrac{3}{4}\dfrac{f}{B}\right), 0 < \dfrac{f}{B} < \dfrac{4}{3}$	≈ 1	0.88
$\dfrac{1.5}{B}\left(1 - \dfrac{3}{2}\dfrac{\|f\|}{B}\right), \dfrac{\|f\|}{B} < \dfrac{2}{3}$	≈ 1	0.88
$\dfrac{2}{B}\exp - \dfrac{2f}{B}, f > 0$	0.76	0.75
$\dfrac{2}{B}\exp - 4\dfrac{\|f\|}{B}$	0.76	0.75

Table 6.2 suggests that ε_∞ often provides a reasonable estimate of the true efficiency ε_t. Moreover even when ε_∞ is much less than ε_t, the asymptotic performance that results from setting μ/B equal to 3 may be acceptable. This choice of μ/B is that which results if the implicit diversity is estimated to be BT and the explicit diversity then optimized.

The above results show that the numerical value of E_b in (6.56b) in general depends on $\sigma(f)$. However as we now show, the minimum value of β for which E_b is positive does not depend on $\sigma(f)$ as long as μ is chosen appropriately. This (universal) value of β is just that which results from an optimized equal-strength diversity system. Thus the capacity of a fading dispersive channel is independent of the channel scattering function and equals $(P/N_0)\log_2 e$—the capacity of the infinite-bandwidth nonfading nondispersive channel that possesses the same power-to-noise ratio P/N_0. This conclusion is a special application of a more general result [13].

To establish the above conclusion, we first note, from (5.14c), that E, and hence E_b, is positive if

$$\frac{R}{C} < \frac{\partial E(\rho)}{\partial \rho}\bigg|_{\rho=0}.$$

Equivalently they are positive if β exceeds $\beta_{\text{cap}}(\mu, T)$, where

$$\beta_{\text{cap}}(\mu, T) = (\ln 2)\left[\frac{\partial E(\rho)}{\partial \rho}\bigg|_{\rho=0}\right]^{-1}. \tag{6.64}$$

Equations 6.57*a* and 6.64 imply that

$$\lim_{T \to \infty} \beta_{\mathrm{cap}}(\mu, T) = (\ln 2)\left\{1 - \frac{1}{\mu} \iint \ln\left[1 + \mu\sigma(f)|u_o(t)|^2\right]\right\}^{-1}, \qquad (6.65)$$

or upon choosing $u_o(t)$ to equal one on a unit interval and zero elsewhere,

$$\lim_{T \to \infty} \beta_{\mathrm{cap}}(\mu, T) = (\ln 2)\left\{1 - \frac{1}{\mu} \int \ln\left[1 + \mu\sigma(f)\right] df\right\}^{-1}. \qquad (6.66)$$

To proceed, we break the integral over f into two parts: one part including those values of f for which $\sigma(f)$ exceeds a parameter δ, and the other part including the remaining values. Since $\sigma(f)$ is absolutely integrable, the first set is of finite "length," or measure, and we denote this length by l. By the concavity of the logarithm, the first integral is bounded by

$$\int_{\sigma(f)>\delta} \ln\left[1 + \mu\sigma(f)\right] df \leq l \ln\left[1 + \frac{\mu}{l}\int_{\sigma(f)>\delta}\sigma(f)\,df\right] \leq l \ln\left(1 + \frac{\mu}{l}\right).$$

By the inequality

$$\ln\left[1 + \mu\sigma(f)\right] \leq \mu\sigma(f),$$

the second integral is bounded by

$$\int_{\sigma(f)\leq\delta} \ln\left[1 + \mu\sigma(f)\right] df \leq \mu \int_{\sigma(f)\leq\delta}\sigma(f)\,df.$$

Consequently for any fixed value of δ, and hence of l,

$$\lim_{T \to \infty} \beta_{\mathrm{cap}}(\mu, T) \leq (\ln 2)\left[1 - \int_{\sigma(f)\leq\delta}\sigma(f)\,df - \frac{l}{\mu}\ln\left(1 + \frac{\mu}{l}\right)\right]^{-1}, \qquad (6.67)$$

or upon letting μ tend to infinity,

$$\lim_{\mu \to \infty}\lim_{T \to \infty} \beta_{\mathrm{cap}}(\mu, T) \leq (\ln 2)\left[1 - \int_{\sigma(f)\leq\delta}\sigma(f)\,df\right]^{-1}. \qquad (6.68)$$

Finally we make the parameter δ arbitrarily small. Since $\sigma(f)$ is nonnegative and integrable over the whole line, the integral over the set such that $\sigma(f) \leq \delta$ vanishes as δ tends to zero. Therefore

$$\beta_{\mathrm{cap}} = \lim_{\delta \to 0}\lim_{\mu \to \infty}\lim_{T \to \infty} \beta_{\mathrm{cap}}(\mu, T) \leq \ln 2 \qquad (6.69)$$

that is, for the $u_o(t)$ we are considering and for sufficiently large values of T and μ, the value of E_b is positive if β exceeds a number β_{cap} that is no larger than $\ln 2$. On the other hand, $\ln 2$ is the value of β for which E_b vanishes in an optimized equal-strength diversity system. Thus β_{cap} must in fact equal $\ln 2$, as was to be proved.

Related Results: We now establish an efficiency bound whose validity is not restricted to the asymptotic situation. This result is of value in estimating the performance that can be realized with realistic waveform durations. We also present a related asymptotic result that is required in Chapter 7. It is convenient to introduce these results in reverse order.

We previously set D_i equal to b^3/d^2 in (6.21c) so as to obtain the asymptotic results. We now set D_i equal to b/λ^2, where λ is an upper bound to all of the eigenvalues λ_i. This is consistent with the constraint of (6.21c) because

$$d = \sum \lambda_i^3 \le \lambda \sum \lambda_i^2 = \lambda b, \tag{6.70a}$$

if

$$\lambda \ge \lambda_i, \qquad i = 1, \dots . \tag{6.70b}$$

With this choice of D_i the expression (6.21b) for ε becomes

$$\varepsilon = \frac{b}{\lambda}. \tag{6.71}$$

It is clear from (6.71) that the sharpest bound is obtained by setting λ equal to λ_0, the dominant eigenvalue. This choice is appropriate in the asymptotic situation we are considering because λ_0 can then be evaluated. Of course in this asymptotic situation the bound of (6.46) is stronger than that of (6.71). However the asymptotic behavior of (6.71) with λ set equal to λ_0 is of interest in Chapter 7, and a knowledge of this asymptotic behavior helps us judge the strength of our subsequent nonasymptotic result in which λ does not equal λ_0.

It is shown in Appendix 4 that the asymptote of the dominant eigenvalue λ_0 is given by the expression

$$\lim_{T \to \infty} T\lambda_0 = |u_m|^2 \sigma_m, \tag{6.72}$$

where

$$|u_m|^2 = \max_t |u_0(t)|^2 \tag{6.73a}$$

and

$$\sigma_m = \max_f \sigma(f). \tag{6.73b}$$

This equation, in conjunction with (6.44) and (6.45), yields the following asymptotic expression for ε, as defined by (6.71) with λ set equal to λ_0:

$$\lim_{T \to \infty} \varepsilon = \varepsilon'_\infty, \tag{6.74a}$$

where

$$\varepsilon'_\infty = \frac{1}{B|u_m|^2 \sigma_m} \tag{6.74b}$$

and as before

$$B = \left\{ \int [\sigma(f)]^2 \, df \right\}^{-1}. \tag{6.74c}$$

It is easy to show that the right member of (6.74a) is maximized by taking $u_0(t)$ to be the rectangular pulse of (6.49). The resulting value of ε'_∞ is

$$\varepsilon'_\infty = \frac{1}{B\sigma_m}. \tag{6.75}$$

This asymptotic value is compared with that of (6.52) for several scattering functions in Table 6.3. It is clear from the table that the difference is not too

Table 6.3 Typical Values of ε_∞ and ε'_∞ for Highly Frequency Dispersive Systems

$\sigma(f)$	ε_∞	ε'_∞
$B^{-1}, \ \lvert f \rvert < \dfrac{B}{2}$	1	1
$\dfrac{2}{B} \exp -4 \dfrac{\lvert f \rvert}{B}$	0.75	0.5
$\dfrac{2}{B} \exp -2 \dfrac{B}{f}, f \geq 0$	0.75	0.5
$\dfrac{\sqrt{2}}{B} \exp -\dfrac{\pi}{2} \left(\dfrac{f}{B}\right)^2, f \geq 0$	0.866	0.707
$\dfrac{2}{B} \left[1 + \left(\dfrac{2\pi f}{B}\right)^2\right]^{-1}$	0.66	0.5
$\dfrac{1.5}{B} \left(1 - \dfrac{3}{4}\dfrac{f}{B}\right), 0 < \dfrac{f}{B} < \dfrac{4}{3}$	0.88	0.66

great. Of course in the asymptotic situation, it is always preferable to employ (6.52) in lieu of (6.75). However we next use (6.71) to establish a result that is valid for all values of T, and it is comforting to know that this equation is capable of yielding results that compare favorably with those of (6.52).

Usually the value of λ_0 can be determined easily only in the asymptotic situation considered above. Therefore if we are to obtain any relatively simple bound to the performance for finite values of T, we must take λ to be larger than λ_0. The choice of λ involves a compromise between the strength of the result and the ease of its application. The following choice appears to be a reasonable one. That it is admissible is shown in Appendix 4.

We define the value of λ by the expression

$$\lambda = \max_{f',r'} \iint \sigma(r,f)|\theta(r-r',f-f')|\, dr\, df, \tag{6.76}$$

where $\theta(r,f)$ is defined by (6.9).

It is satisfying that (6.76) provides an asymptotically correct bound to λ_0 in at least two situations of interest. Specifically if

$$u(t) = \frac{1}{\sqrt{T}}\, 2^{1/4} \exp - \pi \left(\frac{t}{T}\right)^2 \tag{6.79a}$$

or if

$$u(t) = \sqrt{T}\, \frac{\sin \pi(t/T)}{\pi t}, \tag{6.77b}$$

the value of λ converges with increasing T to λ_0, as computed from (6.72). The truth of this assertion is easily demonstrated by direct substitution of (6.77a) or (6.77b) into (6.72) and (6.76). Unfortunately the expression of (6.76) does not yield a finite result for the rectangular waveform of (6.51).

Although it is sometimes laborious, (6.71) with λ defined by (6.76) can be evaluated for many scattering functions and modulation waveforms. For example, if $u(t)$ is given by (6.77a) and if the scattering function is

$$\sigma(r,f) = \frac{2}{BL} \exp - 2\pi \left[\left(\frac{r}{L}\right)^2 + \left(\frac{f}{B}\right)^2\right],$$

we obtain from (6.30), (6.31), and (6.76)

$$|\theta(\tau,f)| = \exp - \frac{\pi}{2}\left[\left(\frac{\tau}{T}\right)^2 + (fT)^2\right],$$

$$b = \left[1 + (BL)^2 + (BT)^2 + \left(\frac{L}{T}\right)^2\right]^{-1/2},$$

and

$$\lambda = \left\{\left[1 + \left(\frac{L}{2T}\right)^2\right]\left[1 + \left(\frac{BT}{2}\right)^2\right]\right\}^{-1/2}.$$

Thus (6.71) yields, after some manipulation,

$$\varepsilon = \left\{\frac{1}{4} + \frac{3}{4}\left[1 + (BL)^2 + (BT)^2 + \left(\frac{L}{T}\right)^2\right]^{-1}\right\}^{1/2}. \tag{6.78}$$

It is worth noting that the result of (6.78) approaches $\frac{1}{2}$ rather quickly, the rate of approach being roughly $(1/BT)^2$. Thus as in the discussion of (6.55), it appears that BT need not be too large, nor L/T too small, before the asymptotic situation is essentially realized. Of course, the asymptotes in these two situations differ because different values of D_i are employed.

Highly Time Dispersive Systems

We say that a system is highly time dispersive if LW is much greater than unity whereas BT is much less than unity. The importance of this mode of operation is analogous to that of highly frequency dispersive systems in that it may simplify the attainment of performance levels that otherwise are difficult, or impossible, to achieve.

The limiting behavior of highly time dispersive systems can be determined easily by utilizing the equivalent channel concepts discussed in Section 2.6. That discussion assures us that the system employing the scattering function

$$\tilde{\sigma}(r, f) = g(r, f) \tag{6.79a}$$

in conjunction with a modulation whose Fourier transform is $U(f)$ has the same performance characteristics as the system employing the scattering function

$$\sigma(r, f) = g(f, -r) \tag{6.79b}$$

in conjunction with the complex (time) waveform $U(t)$. Thus the entire discussion of highly frequency dispersive systems can be transformed into a discussion of highly time dispersive systems by replacing T with W, $u(t)$ with $U(f)$ and the Doppler scattering function $\sigma(f)$ with the range-delay scattering function $\sigma(r)$.

More generally, we can consider a complex modulation envelope of the form

$$\tilde{u}(t) = \left(\frac{a}{T}\right)^{1/2} \iint f_0\left(\frac{y}{T}\right) \exp\ -j\pi(kt^2 + 2tx + cx^2 - 2axy)\ dx\ dy, \tag{6.80a}$$

where $f_0(t)$ is a unit-norm unit-time-duration waveform and a, k, and c are fixed numbers. If this waveform is used in conjunction with a scattering function

$$\tilde{\sigma}(r, f) = g(r, f), \tag{6.80b}$$

the resulting system performs as though the scattering function is

$$\sigma(r, f) = g\left(ar - \frac{cf}{a}, akr + \frac{(1 - kc)}{a}f\right), \tag{6.81a}$$

the modulating waveform is

$$u(t) = \frac{1}{\sqrt{T}} f_0\left(\frac{t}{T}\right), \tag{6.81b}$$

and the Doppler scattering function is

$$\sigma(f) = \int g\left(ar - \frac{cf}{a}, \, akr + \frac{1 - kc}{a}f\right) dr. \tag{6.81c}$$

In particular, if T tends to infinity, the original channel, employed with the waveform of (6.80a), performs as though it is a highly frequency dispersive system with the frequency scattering function of (6.81c) and the modulation envelope of (6.81b). The performance of highly time dispersive systems can be extracted from those equations by taking

$$c = -a,$$

$$k = \frac{-1}{a}$$

and then letting a approach zero; $\sigma(f)$ is then the range-delay scattering function of the original channel and $(1/\sqrt{T})f_o(\cdot)$ is the Fourier transform of the original modulation $\tilde{u}(t)$.

6.6   HIGHLY DOUBLY DISPERSIVE CHANNELS

Roughly stated, a highly doubly dispersive channel is one whose scattering function is very wide in both range delay and Doppler spread. This implies that the spread factor S is quite large. It also implies, with some qualifications, that the two-frequency correlation function is very "narrow" in both its dimensions. We first consider channels that are, in a limiting sense, highly doubly dispersive. We then discuss the factors that limit our ability to make a channel "appear" to be highly doubly dispersive. Finally we consider the performance that can be achieved by using waveforms that possess a high time-bandwidth product. The principal conclusion is that a channel that is sufficiently dispersive can be made to perform as an equal-strength diversity system by resorting to modulating waveforms with high time-bandwidth products.

We measure the performance of systems by the quantities ε and D of (6.21) with D_i set equal to b^3/d^2. Thus

$$\varepsilon = \frac{b^2}{d}, \tag{6.82a}$$

$$D_i = \frac{b^3}{d^2}, \tag{6.82b}$$

where b and d are defined by (6.19). As in previous discussions we interpret ε as the efficiency relative to an optimized equal-strength diversity system.

The Limiting Channel

We first consider the performance of a system with a fixed modulation as the channel becomes more and more dispersive in both dimensions. Precisely stated, we suppose that the scattering function is of the form

$$\sigma(r, f) = \left(\frac{1}{a}\right)^2 \sigma_o\left(\frac{r}{a}, \frac{f}{a}\right) \tag{6.83}$$

and then let the value of a tend to infinity. In this expression $\sigma_o(r, f)$ is a scattering function with unit spread; that is,

$$\iint \sigma_o(r, f) \, dr \, df = 1, \tag{6.84a}$$

$$\iint [\sigma_o(r, f)]^2 \, dr \, df = 1. \tag{6.84b}$$

Thus the spread S of $\sigma(r, f)$ is a^2, and S tends to infinity with increasing a.

The asymptotic expressions for b and d in the limit of increasing S are developed in Appendix 4. Upon introducing these expressions into (6.82) and passing to the limit, we obtain

$$\lim_{s \to \infty} \varepsilon = \left\{ \iint [\sigma_o(r, f)]^3 \, dr \, df \right\}^{-1} \tag{6.85a}$$

and

$$\lim_{s \to \infty} \frac{D_i}{S} = \lim_{s \to \infty} \varepsilon^2. \tag{6.85b}$$

At this juncture several comments are in order. First, the validity of (6.85) has only been established subject to the constraint that S approaches infinity in the manner prescribed by (6.83). However it is probably valid under more general conditions. Second, the result of (6.85) provides additional proof that channel overspreading need not preclude the realization of the optimum performance levels of Chapter 5. In fact, if the channel is sufficiently spread the efficiency of (6.85) is realized no matter what modulating waveform is employed—provided that the proper amount of explicit diversity is employed. Thus, for example, the channel whose scattering function is

$$\sigma(r, f) = S^{-1}, \qquad r^2 + f^2 \leq \frac{S}{\pi},$$

$$\sigma(r, f) = 0, \qquad r^2 + f^2 > \frac{S}{\pi},$$

can be made to perform as an optimized equal-strength diversity system provided that S is sufficiently large.

The approximate diversity estimate of (6.1) and the conclusions of subsequent sections all suggest that the gross characteristics of a system's operation are determined by the magnitudes of BT and LW. For example, if both of these quantities are small, the system behaves as though the channel is only slightly dispersive; whereas if BT is large and LW is small, the system behaves as though the channel is highly dispersive in frequency and nondispersive in time. Thus we might conclude that a system performs as though the channel is highly doubly dispersive if both BT and LW are large.

Unfortunately the validity of (6.85) is not even approximately controlled by BT and LW. In fact it appears that its validity hinges more on the magnitudes of B/W and L/T than on the magnitudes of BT and LW. Since (6.85) provides no insight into the performance than can be achieved by increasing BT and LW for a given channel, we next examine the performance levels that can be realized by employing high TW waveforms in the modulator.

High TW Modulation

As in some other communication problems, it is easier to establish a level of performance that can be achieved with high TW waveforms than it is to find a waveform that attains this performance. In particular, such a level can be established by determining the average performance over an ensemble of waveforms. This technique, which was introduced by Shannon, is the one we employ here [14–16].

Let us consider a set $\{u(t)\}$ of waveforms that might be used for the complex modulation envelopes. A value of ε and D_i as computed from (6.82) can be associated with each of the waveforms. We are free to employ any desired amount of explicit diversity with each waveform, and we suppose that this amount is optimized, as described in Section 6.3, so that ε is a bound to the efficiency of the resulting system measured relative to an optimized equal-strength system.

For our present purposes it is expedient to relax the requirement that the waveforms be of unit norm and to account for this normalization in the expression for ε. It is easy to verify that this can be done by dividing the right member of (6.82a) by the norm of $u(t)$; that is,

$$\varepsilon_u \geq \frac{b_u^{\,2}}{d_u E_u}, \tag{6.86}$$

where

$$E_u = \int |u(t)|^2 \, dt. \tag{6.87}$$

The subscript u has been added to ε_u, b and d to emphasize the dependence of these quantities upon $u(t)$.

We next seek a lower bound to the efficiency that assuredly can be achieved by at least one waveform in the set. To this end we assign a probability measure to the set of waveforms and then consider the average values of ε_u, b_u, d_u, and E_u over the ensemble of waveforms generated by this measure.

Equation 6.86 and the nonnegativeness of the quantities involved imply that

$$\overline{(\varepsilon_u E_u d_u)^{1/2}} \geq \overline{b_u}, \tag{6.88a}$$

where the bar denotes the ensemble average. Also

$$\overline{(\varepsilon_u E_u d_u)^{1/2}} \leq \sqrt{\varepsilon_o} \, \overline{(E_u d_u)^{1/2}} \leq (\varepsilon_o \overline{E_u} \, \overline{d_u})^{1/2} \tag{6.88b}$$

where ε_o denotes the maximum value of the ε_u. The first inequality follows from the definition of ε_o, the second from the properties of moments. Equations 6.88a and b yield

$$\varepsilon_o \geq \frac{(\overline{b_u})^2}{\overline{d_u}\,\overline{E_u}}. \tag{6.89}$$

That is, at least one of the waveforms in the ensemble possesses an efficiency that equals or exceeds the right member of (6.89).

It now remains to choose the ensemble of waveforms and to evaluate the averages in question. The choice is constrained by the feasibility of the subsequent evaluation, one of the few suitable choices being the Gaussian random process. In the interests of simplicity we restrict our attention to the (complex) Gaussian random process for which

$$\overline{u(t)} = 0, \tag{6.90a}$$

$$\overline{u(t)u(\tau)} = 0, \tag{6.90b}$$

and

$$\overline{u(t)u^*(\tau)} = \frac{\sqrt{2}}{T} \exp - \frac{\pi}{T^2}\left[t^2 + \tau^2 + \frac{(TW)^2}{2}(t - \tau)^2\right]. \tag{6.90c}$$

This is probably not the best possible choice, but it leads to more manageable results than several others do. The same desire for simplicity leads us to restrict the following discussion to the asymptotic behavior of ε_o as both T and W increase without limit.

The derivation of the limiting values of $\overline{E_u}$, $\overline{b_u}$, and $\overline{d_u}$, for the correlation function of (6.90c), is presented in Appendix 4 and the results are summarized in Theorem 1 of that Appendix. Since the strongest of these limiting expressions is rather cumbersome, we confine our discussion to the simpler but weaker ones. The use of the stronger results is, however, straightforward.

According to Theorem 1 of Appendix 4,

$$\bar{E}_u = 1, \tag{6.91a}$$

$$\lim TW\bar{b}_u = 1 + S^{-1}, \tag{6.91b}$$

and

$$\lim (TW)^2 \bar{d}_u \leq \tfrac{4}{3}[2 + (3 + \sigma)S^{-1}], \tag{6.91c}$$

where

$$S^{-1} = \iint [\sigma(r, f)]^2 \, dr \, df \tag{6.92a}$$

and

$$\sigma = \max_{r, f} \sigma(r, f). \tag{6.92b}$$

Thus by virtue of (6.89),

$$\varepsilon_o \geq \frac{3}{4} \frac{[1 + S^{-1}]^2}{2 + (3 + \sigma)S^{-1}}, \tag{6.93}$$

and at least one sample function of the Gaussian random process described by (6.90) yields a value of ε that exceeds the right member of this equation.

Several important conclusions follow immediately from (6.93). First, if σ is constrained to be less than a fixed number and if S becomes exceedingly large, the right member of the equation tends to 0.375. That is, no matter what the shape of the scattering function, an efficiency of at least 0.375 can be realized provided that S is sufficiently large and that σ is not too large.

Second, if we fix σ and seek the value S_m of S that minimizes the right member of (6.93), we find that

$$S_m = \infty, \qquad \text{if } \sigma \leq 1,$$

$$S_m = \frac{\sigma + 3}{\sigma - 1}, \qquad \text{otherwise.}$$

Substituting these values into the expression for ε_o leads us to the inequalities (6.94). The right members of these expressions are lower bounds to the efficiency that can assuredly be achieved by at least one waveform in the ensemble no matter what the value of S.

$$\varepsilon_o \geq \tfrac{3}{8}, \qquad \text{if } \sigma \leq 1, \tag{6.94a}$$

$$\varepsilon_o \geq 3 \frac{(1 + \sigma)}{(3 + \sigma)^2}, \qquad \text{if } \sigma > 1. \tag{6.94b}$$

Thus an efficiency of at least 0.375 can be realized if σ is no greater than unity.

Unfortunately as σ increases without limit, the lower bound of (6.94*b*) vanishes as does the stronger bound of (6.93) (if S is fixed). Since scattering functions exist with arbitrary values of S and arbitrarily large values of σ, the result of (6.94*b*) fails to provide any universal positive lower bound to the attainable efficiency. This failure persists when the stronger results of Theorem 1, Appendix 4, are employed.

6.7 LIMITATIONS OF HIGHLY DISPERSIVE CHANNELS

All of the preceding discussion reinforces the conclusion that large spread factors need not preclude the attainment of nearly optimum performance levels. On the other hand, some problems do arise as the channel spreading increases. We now briefly discuss some of these problems in order to gain a better understanding of the true limitations of overspread channels.

One problem is that large amounts of implicit effective diversity are encountered with highly overspread channels. In order to attain the optimum performance levels with this diversity, we must employ a high energy-to-noise ratio per transmission. This, in turn, requires a very large alphabet size.

To quantify these comments, we resort to the bound of (5.86). Specifically if a system is to perform as an optimized system, it is necessary that

$$\alpha b \geq \alpha_p{}^\circ, \tag{6.95}$$

where $\alpha_p{}^\circ$ is the optimum energy-to-noise ratio per diversity path for the given value of β. On the other hand, by virtue of (6.8),

$$b = \iint |\mathscr{R}(x, y)\theta(y, x)|^2 \, dx \, dy \leq \iint |\mathscr{R}(x, y)|^2 \, dx \, dy = S^{-1}. \tag{6.96}$$

Thus the optimum performance can be achieved only if

$$\alpha \geq S\alpha_p{}^\circ; \tag{6.97}$$

that is, the value of α must increase linearly with S if the optimum performance levels are to be attained. For a fixed value of β, this implies that the alphabet size must increase exponentially with S.

Another difficulty encountered with highly overspread channels is that high peak powers are often required. That is, not only must the total energy-to-noise ratio be large but it must also be concentrated, to some degree, in both time and frequency. To demonstrate this, we again employ (6.95) and (6.96).

By virtue of (6.96)

$$b \leq \iint \left[\max_{x'} |\mathscr{R}(x', y)|^2 \right] |\theta(y, x)|^2 \, dx \, dy.$$

Moreover

$$\int |\theta(y, x)|^2 \, dx \leq \int |\theta(0, x)|^2 \, dx = \int |u(t)|^4 \, dt \leq \max_t |u(t)|^2.$$

Therefore

$$b \leq \left[\int |\mathscr{R}_m(y)|^2 \, dy \right] \max_t |u(t)|^2, \qquad (6.98a)$$

where

$$|\mathscr{R}_m(y)| = \max_{x'} |\mathscr{R}(x', y)| \qquad (6.98b)$$

It follows from (6.98) that (6.95) can be satisfied only if

$$\alpha \max_t |u(t)|^2 > \alpha_p{}^o \left[\int |\mathscr{R}_m(y)|^2 \, dy \right]^{-1}.$$

However, the left member of the equation is just the peak-power level at the transmitter, and so the peak transmitted power must increase linearly with

$$\left[\int |\mathscr{R}_m(y)|^2 \, dy \right]^{-1}.$$

This quantity is often approximately equal to the Doppler spread, B. For example it equals B when

$$\mathscr{R}(x, y) = \mathscr{R}(x)\mathscr{R}(y). \qquad (6.99)$$

Thus to a first approximation, we may say that the peak transmitter power must exceed $\alpha_p{}^o B$ if the optimum performance levels are to be attained. Clearly the required peak power increases as either B or $\alpha_p{}^o$ increases.

A similar analysis shows that the optimum performance levels cannot be achieved unless

$$\alpha \max_f |U(f)|^2 \geq \alpha_p{}^o \left[\int |\hat{\mathscr{R}}_m(x)|^2 \, dx \right]^{-1}, \qquad (6.100a)$$

where

$$|\hat{\mathscr{R}}_m(x)| = \max_y |\mathscr{R}(x, y)| \qquad (6.100b)$$

and where $U(f)$ is the Fourier transform of $u(t)$. This result is summarized in the statement that the peak value of the transmitted power density spectrum must exceed the right member of (6.100a) if the optimum performance levels are to be realized. In particular, if $\mathscr{R}(x, y)$ is of the form given by (6.99) the right member of (6.100a) reduces to $\alpha_p{}^o L$, and the peak value of the transmitted power density spectrum must increase linearly with L.

We have approached the waveform design problem from several vantage points in an attempt to gain an understanding of the important issues. Although we have not been able to solve the problem in any complete sense, all of our conclusions suggest that relatively simple diversity estimates can be used to select a good waveform and that these waveforms yield performance levels that are often comparable to the optimum levels. Other approaches to the waveform design problem have led to similar conclusions [6, 9, 17, 18].

REFERENCES

[1] I. L. Lebow, P. R. Drouilhet, N. L. Daggett, J. N. Harris and F. Nagy, "The West Ford Belt as a Communications Medium." *Proc. IEEE*, p. 543, May 1964.

[2] R. S. Kennedy and I. L. Lebow, "Signal Design for Dispersive Channels." *IEEE Spectrum*, pp. 231–237, March 1964.

[3] R. Courant and D. Hilbert, *Methods of Mathematical Physics*. New York: Interscience 1953, p. 133.

[4] F. G. Tricomi, *Integral Equations*. New York: Interscience, 1957, pp. 118–124.

[5] F. Smithies, *Integral Equations*. London: Cambridge University Press, 1958, pp. 130–135.

[6] H. L. Van Trees, *Detection, Estimation, and Modulation Theory*. New York: Wiley, 1970, **2**, Chapter 4.

[7] R. Price and P. Green, "Signal Processing in Radar Astronomy-Communication via Fluctuating Multipath Media," Lincoln Laboratory, MIT, Tech. Rept. 234, pp. C78–C85, DDC No. 246782, 1960.

[8] G. H. Hardy, J. E. Littlewood, G. Polya, *Inequalities* (2nd ed.). London: Cambridge University Press, 1952, p. 132.

[9] N. J. Bershad, "On the Optimum Design of Multipath Signals." *IEEE Trans. Inform. Theory*, pp. 389, October 1964.

[10] R. Courant and D. Hilbert, *Methods of Mathematical Physics*. New York: Interscience, 1953, p. 65.

[11] J. N. Pierce, "Error Probabilities for a Certain Spread Channel." *IEEE Trans. Commun. Systems*. pp. 120–122, March 1964.

[12] S. Halme, "Efficient Optical Communication Through a Turbulent Atmosphere." M.I.T., Research Laboratory of Electronics, Q.P.R. No. 91, October 15, 1968.

[13] J. N. Pierce, "Ultimate Performance of M-ary Transmissions on Fading Channels." *IEEE Trans. Inform. Theory*, pp. 2–5, January 1966.

[14] C. E. Shannon and W. Weaver, *The Mathematical Theory of Communication*. Urbana: University of Illinois Press, 1949.

[15] R. M. Fano, *Transmission of Information*. New York: Wiley, and M.I.T. Press, pp. 231–240, 1961.

[16] R. G. Gallager, *Information Theory and Reliable Communication*. New York: Wiley, 1968, Chapter 5.

[17] N. J. Bershad, "Optimum Binary FSK for Transmitted Reference Systems Over Rayleigh Fading Channels." *IEEE Trans. Commun. Technology*, pp. 784–790, December 1966.

[18] R. F. Daly, "Signal Design for Efficient Detection in Randomly Dispersive Media," Stanford Research Institute Report, Project 186531-144, Electronics and Radio Sciences, Communication Laboratory.

Chapter 7

Demodulators

Thus far it has been assumed that an optimum demodulator is employed in the communication system under consideration. Although the mathematical operations that such a demodulator must perform are specified in Chapter 4, little insight has been gained into the functional structure that the demodulator may possess. One of the objectives of this chapter is to establish some more useful descriptions of optimum demodulators. We conclude that an optimum demodulator is often exceedingly difficult to build. Therefore we also consider suboptimum demodulator structures that circumvent some of the difficulties inherent in optimum systems.

7.1 OPTIMUM DEMODULATORS

We first recall from Chapter 4 that the optimum demodulator consists, in essence, of m distinct branch demodulators, where m is the number of modulator waveforms. We also note that each branch demodulator in a D-fold explicit diversity system is a simple combination of D elementary demodulators, each elementary demodulator being optimum for one of the D parts of the composite waveform. Thus it suffices to consider the structure of the demodulators for systems that do not employ explicit diversity.

Equation 4.17 implies that the jth branch demodulator should evaluate the quantity

$$z_k = \alpha \sum_i \frac{\alpha \lambda_i}{1 + \alpha \lambda_i} |z_{ik}|^2, \qquad (7.1a)$$

where

$$z_{ik} = \frac{1}{\sqrt{E_r}} \int r(t)\varphi_i^*(t) \exp - j\tilde{\omega}_k t \, dt \qquad (7.1b)$$

and

$$\tilde{\omega}_k = 2\pi[f_o + (k-1)\Delta]. \qquad (7.1c)$$

In these expressions, $r(t)$ denotes the received waveform, ω_k the "carrier frequency" of the kth transmitted waveform, E_r the average received signal energy, α the average received energy-to-noise ratio, and the $\varphi_i(t)$ and λ_i denote, respectively, the eigenfunctions and eigenvalues of the complex correlation function $R(t, \tau)$ given by (4.4). A factor of $\sqrt{2}$ has been suppressed in the definition of the z_{ij}, that is, the r_{ij} of (4.17) equal $\sqrt{2}z_{ij}$.

Functional Description

Our immediate objective is to translate the right member of (7.1a) into a functional expression that portrays the character of the operation to be performed upon $r(t)$. To achieve this objective, we introduce (7.1b) into (7.1a) and interchange the order of summation and integration. This interchange is justified by our assumption that the number of scatterers is finite; the correlation function is then degenerate and has a finite number of positive eigenvalues.

Upon defining the function

$$h_k(t, \tau) = \frac{\alpha^2}{E_r} \, h(t, \tau) \exp j\tilde{\omega}_k(t - \tau), \tag{7.2a}$$

where

$$h(t, \tau) = \sum_i \frac{\lambda_i}{1 + \alpha\lambda_i} \, \varphi_i(i)\varphi_i^*(\tau), \tag{7.2b}$$

we obtain

$$z_k = \iint r(t)h_k(t, \tau)r(\tau) \, dt \, d\tau. \tag{7.3}$$

Equation 7.3 implies that the operation of each branch demodulator may be described as the evaluation of a double integral involving $r(t)$ and $r(\tau)$. This interpretation and others derived from it are quite useful. We discuss them later after noting some equivalent specifications of $h(t, \tau)$, which is called the *demodulator kernel*.

The defining expression for $h(t, \tau)$ is difficult to evaluate because it involves the eigenfunctions $\varphi_i(t)$. There are two other equivalent expressions that do not depend explicitly upon the eigenfunctions. These expressions can be derived directly from (7.2b); they are also consequences of the fact that $-h(t, \tau)$ is the Fredholm resolvent of $R(t, \tau)$ [1–3].

The first alternate description provides a means of verifying whether or not any given function of two variables is the demodulator kernel for a specific $R(t, \tau)$. In particular $h(t, \tau)$ is known to be the unique solution of the integral equation

$$-h(t, \tau) + R(t, \tau) = \alpha \int_{-\infty}^{\infty} h(t, x)R(x, \tau) \, dx. \tag{7.4}$$

The remaining alternate provides an explicit means of evaluating $h(t, \tau)$ for values of α that are not too large. Specifically

$$h(t, \tau) = \sum_{i=1}^{\infty} (-\alpha)^{i-1} \tilde{R}_i(t, \tau), \qquad (7.5)$$

where

$$\tilde{R}_1(t, \tau) = R(t, \tau) \qquad (7.6a)$$

and

$$\tilde{R}_i(t, \tau) = \int \tilde{R}_{i-1}(t, x) \tilde{R}_1(x, \tau) \, dx, \qquad \text{for} \quad i > 1. \qquad (7.6b)$$

This series converges in general only for values of α that do not exceed the reciprocal of the dominant eigenvalue. We make little use of (7.4) but subsequently employ (7.5) to suggest some useful suboptimum receiver structures. Additional insight into the operation of the optimum branch demodulator may be obtained by reformulating the integral of (7.3). These forms, which have been presented by others [4–12], are discussed here to provide a framework for the subsequent discussion of the suboptimum structures.

Structural Forms

Equation 7.2 implies that each of the $h_j(t, \tau)$ possesses conjugate symmetry, that is,

$$h_j(t, \tau) = h_j^*(\tau, t). \qquad (7.7)$$

Therefore the integral of (7.3) may be restated as

$$z_j = 2 \int_{-\infty}^{\infty} r(t) \, dt \int_{-\infty}^{t} [\text{Re } h_j(t, \tau)] \, r(\tau) \, d\tau \qquad (7.8)$$

and we may implement the jth branch demodulator in the form illustrated in Fig. 7.1. That is, to compute z_j, the received waveform is passed through the realizable linear time-varying filter whose impulse response is $2 \text{ Re } h_j(t, \tau)$ and the output of the filter is cross-correlated with its input.

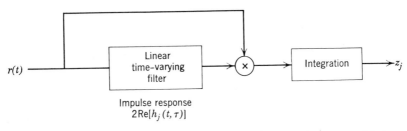

Fig. 7.1 The optimum demodulator implemented as an estimator-correlator.

It can be shown that the filter of Fig. 7.1 yields the minimum rms error estimate of the "signal" component of the noisy waveform existing at the receiver terminals, given that the *j*th waveform is transmitted. Thus the *j*th branch demodulator may be interpreted as a processor that generates a least mean-square error estimate of the signal component of the received waveform, under the hypothesis that the *j*th waveform is transmitted, and then cross-correlates the received waveform with this estimate. This interpretation is conceptually satisfying, but its appeal is diminished by the difficulty of implementing the time-varying filter of Fig. 7.1.

The kernel $h_j(t, \tau)$ can also be expressed in the form

$$h_j(t, \tau) = \int q_j(t, x) q_j^*(x, \tau) \, dx, \tag{7.9}$$

where $q_j(t, x)$ is a Hermitian bounded integrable kernel. The output of the *j*th branch demodulator then may be expressed as

$$z_j = \int_{-\infty}^{\infty} dx \left| \int_{-\infty}^{\infty} r(t) q_j(t, x) \, dt \right|^2. \tag{7.10}$$

This formulation casts the *j*th branch demodulator in the form of a linear filter followed first by a square-law envelope device and then by an integrator.

The preceding descriptions are usually quite difficult to implement because of the time-varying filters involved. It is also difficult to obtain an explicit closed-form expression for the demodulator kernel $h(t, \tau)$. Moreover the design of the optimum demodulator depends on the value of α, the energy-to-noise ratio. This is unsatisfying since α is often either an unknown or a variable. Because of these difficulties, we consider some suboptimum structures. We arrive at these structures by first considering special circumstances wherein the determination or implementation of the optimum demodulator is relatively straightforward. One such structure is the correlation kernel demodulator.

7.2 CORRELATION KERNEL DEMODULATORS

There are two interesting situations in which $h(t, \tau)$ is proportional to the complex correlation function, $R(t, \tau)$.† The less important circumstance occurs when α is exceedingly small. This conclusion follows immediately from the series of (7.5). The second, and much more important, circumstance pertains to the character of the waveform employed.

---

† Since the constant of proportionality does not influence subsequent comparison of the branch demodulator outputs, its value is unimportant and may be ignored.

Specifically $h(t, \tau)$ is proportional to $R(t, \tau)$ if the positive eigenvalues of $R(t, \tau)$ are all equal to each other. This conclusion, which follows readily from (7.2b) in conjunction with the series expansion of (2.42), is quite important because the optimum modulating waveform is one for which the positive eigenvalues of $R(t, \tau)$ are all equal to each other. Of course, we are free to employ $R(t, \tau)$ in lieu of $h(t, \tau)$, be they proportional or not. Moreover if the modulating waveform is a fairly good one, in the sense that the positive eigenvalues are almost equal, the resulting performance loss should be small. This loss is now discussed quantitatively.

The estimate of the loss in performance due to the use of a correlation kernel demodulator is obtained in two steps. In the first step the performance is compared to that of a different system employing the optimum demodulator. In the second step the estimate is reduced to one that relates the performance of the original system to that of an equal-strength diversity system.

Loss Relative to Optimum Demodulators

We first show that the error probability of a suboptimum correlation kernel system with an energy-to-noise ratio α and eigenvalues λ_i is less than the error probability of a system employing the optimum demodulator with an energy-to-noise ratio of

$$\hat{\alpha} = \alpha \frac{1 + \alpha b}{1 + \alpha \lambda} \tag{7.11a}$$

and eigenvalues

$$\hat{\lambda}_i = \lambda_i \frac{1 + \alpha \lambda_i}{1 + \alpha b}. \tag{7.11b}$$

In these expressions λ denotes any upper bound to the λ_i and b is defined as

$$b = \sum_i \lambda_i^2. \tag{7.11c}$$

The validity of the assertion encompassing (7.11) may be established as follows: Equations 7.2a and 7.3, with $R(t, \tau)$ substituted for $h(t, \tau)$, in conjunction with the series expansion of (2.42) for $R(t, \tau)$, imply that the output of the jth branch demodulator may be expressed as

$$z_j = \alpha \sum_i \alpha \lambda_i |z_{ij}|^2, \tag{7.12}$$

where the z_{ij} are defined by (7.1b). By virtue of the decision rule employed, a decision error occurs whenever the output of the "correct" demodulator is exceeded by the outputs of one or more of the "incorrect" demodulators.

Thus if the kth waveform is transmitted, an error occurs if

$$\sum_i \alpha \lambda_i |z_{ik}|^2 < \sum_i \alpha \lambda_i |z_{ij}|^2 \qquad (7.13)$$

for one or more values of j other than k.

A more useful form of (7.13) is

$$\sum_i \lambda_i (1 + \alpha \lambda_i) |x_{ik}|^2 < \sum_i \lambda_i |x_{ij}|^2, \qquad \text{one or more } j \neq k, \qquad (7.14)$$

where the x_{ik} are defined by the expression

$$x_{ik} = z_{ik} \left(\frac{\alpha}{1 + \alpha \lambda_i} \right)^{1/2} \qquad (7.15a)$$

and the x_{ij} by the expression

$$x_{ij} = \sqrt{\alpha} z_{ij}, \qquad j \neq k. \qquad (7.15b)$$

For purposes of comparison we note that, when an optimum receiver is used in conjunction with eigenvalues $\hat{\lambda}_i$ and an energy-to-noise ratio $\hat{\alpha}$, an error occurs if

$$\sum_i \hat{\lambda}_i |\hat{x}_{ik}|^2 < \sum_i \frac{\hat{\lambda}_i}{1 + \hat{\alpha} \hat{\lambda}_i} |\hat{x}_{ij}|^2, \qquad \text{one or more } j \neq k, \qquad (7.16)$$

where the $\hat{x}_{ik}$ and $\hat{x}_{ij}$ are given by (7.15) with $\hat{\alpha}$ and $\hat{\lambda}_i$ replacing α and λ_i. Note that the $\hat{x}_{ik}$ and $\hat{x}_{ij}$ possess the same statistical properties as do the x_{ik} and x_{ij}.

It is clear that an upper bound to the error probability of the given sub-optimum system is provided by the probability that

$$\sum_i \lambda_i (1 + \alpha \lambda_i) |x_{ik}|^2 < \sum_i \tilde{\lambda}_i |x_{ij}|^2, \qquad \text{one or more } j \neq k, \qquad (7.17)$$

where the $\tilde{\lambda}_i$ are any upper bounds to the λ_i. In particular, we may specify $\tilde{\lambda}_i$ by the expression

$$\frac{\lambda_i (1 + \alpha \lambda_i)}{1 + \alpha \lambda_i (1 + \alpha \lambda_i)/(1 + \alpha \lambda)},$$

where λ is any upper bound to the $\tilde{\lambda}_i$. This choice permits us to cast (7.17) in the form of (7.16) with

$$\hat{\lambda}_i = \lambda_i \frac{1 + \alpha \lambda_i}{1 + \alpha b}$$

and

$$\hat{\alpha} = \alpha \frac{1 + \alpha b}{1 + \alpha \lambda}.$$

Thus, as asserted, the error probability of the given suboptimum system is no larger than that of the optimum receiver system described by (7.11).

The estimate for the performance of a correlation kernel receiver provided by (7.11) is of some value, but it is subject to all of the shortcomings encountered in the treatment of optimum receivers. Therefore we further simplify (and weaken) it by an appeal to the results of Chapter 6. Specifically the error probability is further over estimated by that of a *D*-fold equal-strength diversity system; that is, we measure the degradation of the given (correlation kernel) system relative to a system that employs both an optimum receiver and also equal-strength diversity. Thus not all of this loss could be recovered by employing the optimum receiver alone.

Loss Relative to Equal-strength Systems

The bound we seek follows readily from the statements encompassing (7.11) in conjunction with (5.77). The latter equation implies that a system that employs an optimum receiver with given values of $\hat{a}$, $\hat{\lambda}_i$, and v performs at least as well as an equal-strength diversity system with a (total) energy-to-noise ratio of $\varepsilon\hat{a}$ and an energy-to-noise ratio per diversity path of $\hat{a}\sqrt{\hat{b}/D}$, where

$$\varepsilon = \sqrt{D\hat{b}}, \tag{7.18a}$$

$$\hat{b} = \sum_i \hat{\lambda}_i^2, \tag{7.18b}$$

and

$$\hat{a} = \sum_i \hat{\lambda}_i^3. \tag{7.18c}$$

The value of the effective diversity D can be chosen at will provided that it satisfies the constraint

$$D \leq \frac{\hat{b}^3}{\hat{a}^2}. \tag{7.18d}$$

Equations 7.18 imply that the (optimum receiver) system described by (7.11) performs at least as well as an equal-strength diversity system with a total energy-to-noise ratio of $\varepsilon\hat{a}$ and an energy-to-noise ratio per diversity path of $\hat{a}\sqrt{\hat{b}/D}$ where

$$\varepsilon = \left[D \sum_i \lambda_i^2 \frac{(1 + \alpha\lambda_i)^2}{(1 + \alpha b)^2} \right]^{1/2} \tag{7.19a}$$

provided that D satisfies the constraint

$$D \leq \frac{\left[\sum_i \lambda_i^2 (1 + \alpha\lambda_i)^2 \right]^3}{\left[\sum_i \lambda_i^3 (1 + \alpha\lambda_i)^3 \right]^2} \tag{7.19b}$$

As in Chapter 6, the best bound results when (7.19b) is an equality, but it is more convenient to employ a bound that involves only the values of α, b, and λ. Specifically we observe that the right member of (7.19b) exceeds $\hat{b}[(1 + \alpha b)/\lambda(1 + \alpha\lambda)]^2$, where λ is any upper bound to the λ_i, and we set D equal to $\hat{b}[(1 + \alpha b)/\lambda(1 + \alpha\lambda)]^2$. The effective energy-to-noise ratio per diversity path then becomes

$$\alpha_p = \frac{\hat{\alpha}\lambda(1 + \alpha\lambda)}{1 + \alpha b}$$

or by virtue of (7.11),

$$\alpha_p = \alpha\lambda. \tag{7.20}$$

Also, the value of ε becomes

$$\varepsilon = \frac{\hat{b}}{\lambda}\frac{1 + \alpha b}{1 + \alpha\lambda}; \tag{7.21a}$$

or, since $\hat{b}$ exceeds b

$$\varepsilon \geqslant \frac{b}{\lambda}\frac{1 + \alpha b}{1 + \alpha\lambda}. \tag{7.21b}$$

Hence the (total) effective energy-to-noise ratio satisfies the inequality

$$\hat{\alpha}\varepsilon \geqslant \frac{\alpha b}{\lambda}\left(\frac{1 + \alpha b}{1 + \alpha\lambda}\right)^2. \tag{7.22}$$

Equations 7.20 and 7.22 imply that the original system employing a correlation kernel receiver with an energy-to-noise ratio of α performs at least as well as equal-strength diversity system employing the optimum demodulator with an energy-to-noise ratio of $(\alpha b/\lambda)[(1 + \alpha b)/(1 + \alpha\lambda)]^2$ and an energy-to-noise ratio per diversity path of $\alpha\lambda$. Consequently we may interpret $(b/\lambda)[(1 + \alpha b)/(1 + \alpha\lambda)]^2$ as a lower bound to the efficiency of the original system relative to this equal-strength system. We denote this bound to the efficiency by ε_1 to distinguish it from those of Chapter 6 which pertain to optimum demodulators. Thus

$$\varepsilon_1 = \frac{b}{\lambda}\left(\frac{1 + \alpha b}{1 + \alpha\lambda}\right)^2, \tag{7.23}$$

where b is given by (7.11c) and λ is any upper bound to the λ_i.

As in Chapter 6 it is instructive to determine the best performance that can be achieved through the use of explicit diversity. That is, we replace λ by $D_e^{-1}\lambda$ and b by $D_e^{-1}b$, where D_e is the order of explicit diversity employed; we then inquire into the error probability that can be realized by variation of D_e.

When we undertook a similar optimization in Chapter 6, we found that the optimization of D_e reduced to the optimization of the effective energy-to-noise ratio per diversity path in an equal-strength system. The present situation is more complex because ε_1 depends on D_e, and so we should maximize the error-probability exponent with respect to both ε_1 and the effective energy-to-noise ratio per diversity path $\alpha\lambda/D_e$.

In some specific problems it is worthwhile to undertake this maximization, but it is then usually more appropriate to work with one of the sharper results obtained in the course of arriving at (7.23). For simplicity we choose here to seek expressions that can be interpreted in terms of the results of Chapter 5. In particular we arbitrarily choose D_e so that the effective energy-to-noise ratio per diversity path $\alpha\lambda/D_e$ equals $\alpha_p{}^\circ$ as determined from Fig. 5.4. Thus

$$D_e = \frac{\alpha\lambda}{\alpha_p{}^\circ}. \tag{7.24}$$

When the order of explicit diversity is chosen in this manner, the resulting value of ε_1, which we denote by ε_2, is

$$\varepsilon_2 = \frac{b}{\lambda}\left[\frac{1 + (b/\lambda)\alpha_p{}^\circ}{1 + \alpha_p{}^\circ}\right]^2. \tag{7.25}$$

Equation 7.25 provides an estimate of the efficiency of the given system relative to an optimized equal-strength diversity system. Precisely stated, a system employing a correlation kernel receiver, an energy-to-noise ratio per information bit of β, and D_e-fold explicit diversity with D_e specified by (7.24) performs at least as well as an optimized equal-strength system employing an energy-to-noise ratio per information bit of $\beta\varepsilon_2$.

To assign some rough numerical value to (7.25), we set λ equal to the dominant eigenvalue λ_o and consider the highly frequency dispersive mode of operation. The value of b/λ is then equal to ε'_∞ as given by Table 6.3. Since ε'_∞ often exceeds 0.7, ε_2 may be expected to exceed about 0.4 when $\alpha_p{}^\circ$ equals 3 (β exceeds the critical value) and about 0.35 as $\alpha_p{}^\circ$ approaches infinity.

The result of (7.25) is appealing in that it is simple and yet valid for all values of β. These attributes, however, have been obtained at the expense of weakening the results; much stronger results can be obtained by straight-forward applications of the bounds presented in Chapter 4 and in Appendix 3. We do not pursue the development here other than to state (without proof) that, for large values of β, ε_2 may be taken to be b/λ.

We have now established an important situation in which the kernel for the optimum demodulator reduces to the complex correlation function; we have also found that the performance loss attendant to the use of the correlation kernel demodulator is not unreasonable even when it is not optimum. On the other hand, we have not yet shown that the implementation of the correlation kernel demodulator is any simpler than that of the optimum demodulator. Fortunately there are circumstances in which it is simpler, and these circumstances suggest other demodulator structures that are sometimes preferable to the correlation demodulator for reasons of simplicity even when they give rise to a further loss of performance. These special forms occur when the channel appears to be either only slightly dispersive or dispersive in only one dimension.

7.3 SLIGHTLY DISPERSIVE CHANNELS

We recall from (6.24) that the basic complex correlation function for a non-dispersive channel is

$$R(t, \tau) = u(t - r_0)u^*(\tau - r_0) \exp j\omega_d(\tau - t) \qquad (7.26)$$

with

$$\omega_d = 2\pi f_d,$$

where r_0 and f_d are the nominal range delay and Doppler shift associated with the channel. It follows easily from (7.4) and (7.26) that $h(t, \tau)$ is proportional to $R(t, \tau)$ for a nondispersive channel. Thus the output z_i of the (optimum) ith branch demodulator may be expressed in the form

$$z_i = \left| \int r(t)u^*(t - r_0) \exp - j(\tilde{\omega}_i - \omega_d)t \, dt \right|^2, \qquad (7.27)$$

where $\tilde{\omega}_i$ is the "carrier" frequency of the ith modulator waveform (7.1).

Demodulator Structure

The right member of (7.27) can be interpreted in several useful ways [13–16]. All of these interpretations stem from the observations that $|f(t) \exp j\omega_0 t|^2$ is the squared envelope of the time function $\mathrm{Re}\,[f(t) \exp j\omega_0 t]$ and that the quantity $\mathrm{Re}\,[\int r(t)q(\tau - t) \exp j\omega_0(\tau - t) \, dt]$ equals the output, at time τ, of a filter whose impulse response is $\mathrm{Re}\,[q(t) \exp j\omega_0 t]$ and whose input $r(t)$ is real. Thus $|\int r(t)q(\tau - t) \exp j\omega_0(\tau - t) \, dt|^2$ equals the squared envelope of the filter output at time τ. These statements are summarized in Fig. 7.2a.

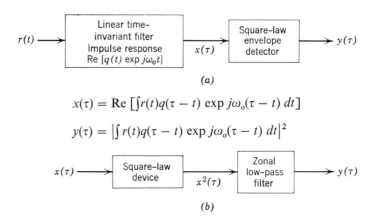

(a)

$$x(\tau) = \text{Re} \left[\int r(t) q(\tau - t) \exp j\omega_o(\tau - t) \, dt \right]$$

$$y(\tau) = \left| \int r(t) q(\tau - t) \exp j\omega_o(\tau - t) \, dt \right|^2$$

(b)

Fig. 7.2 Complex envelope relation.

The envelope detector of Fig. 7.2a has been defined only mathematically, but if either $r(t)$ or $q(t)$ is a narrow-band function with respect to the carrier frequency ω_o, the process of envelope detection and squaring may be approximately realized by the configuration shown in Fig. 7.2b. Specifically a square-law detector followed by a low-pass filter yields a satisfactory approximation of the desired output provided that the filter bandwidth is much greater than the bandwidth of both $r(t)$ and of $q(t)$ but is much less than the carrier frequency [15, 16].

Let us now return to the interpretation of (7.27) and express the right member of that equation in the form

$$z_i = \left| \int r(t) \, q(r_o - t) dt \right|^2, \tag{7.28a}$$

where

$$q(t) = u^*(-t) \exp j(\tilde{\omega}_i - \tilde{\omega}_d)t. \tag{7.28b}$$

The filter whose impulse response is $\text{Re}\,[q(t)]$ is generally referred to as the filter matched to $\text{Re}\,[u(t) \exp j(\tilde{\omega}_i - \omega_d)t]$ or simply as a *matched filter* [17, 18]. Thus the ith branch demodulator consists of a matched filter followed by a square-law envelope detector which is sampled at time r_o. This point of view is emphasized in Fig. 7.3.

Fig. 7.3 Optimum demodulator for nondispersive channel.

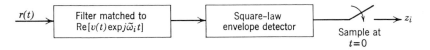

Fig. 7.4 General filter-squarer demodulator; $\int |v(t)|^2 \, dt = 1$.

We recall from Chapter 3 that a given channel appears to be nondispersive for a given modulating waveform if both BT and LW are much less than unity. Of course these conditions are only approximate and presuppose that the modulating waveform and the channel scattering function are concentrated in both time (or range delay) and frequency (or Doppler). However it is true that the optimum branch demodulator can be well approximated by the structure depicted in Fig. 7.3 whenever both BT and LW are sufficiently small. On the other hand, in the interests of system simplicity, we may choose to use this structure even when it is not optimum. Thus it is of interest to determine the performance of that system, be it optimum or not. More generally, we determine the performance of the system shown in Fig. 7.4 for an arbitrary filter response.

Filter-Squarer Performance

We suppose here that the demodulator outputs of the system shown in Fig. 7.4 satisfy the conditional independence conditions summarized in Fig. 4.9. These conditions are approximately satisfied if the frequency offsets $\tilde{\omega}_i - \tilde{\omega}_k$ between the various message waveforms exceed one half the sum of the filter's bandwidth, the transmitted waveform's bandwidth, and the channel Doppler spread. The conditions are precisely satisfied if $v(t)$ $\exp j(\tilde{\omega}_i - \tilde{\omega}_k)t$ is orthogonal to the basic complex correlation function $R(t, \tau)$ for all distinct values of i and k.

When the conditions are satisfied, the analysis of the system is straightforward. Indeed the output of the filter is a Gaussian random process, its envelope is Rayleigh distributed, and its envelope squared is chi squared with two degrees of freedom [19, 20]. Moreover the output statistics of the branch demodulators are identical to those that occur with a constant Rayleigh, or flat-flat, fading channel with an energy-to-noise ratio of $\alpha\tilde{\varepsilon}$, where α is the average received energy-to-noise ratio

$$\tilde{\varepsilon} = \iint v^*(t)R(t, \tau)v(\tau) \, dt \, d\tau \qquad (7.29a)$$

and $R(t, \tau)$ is again the basic complex correlation function of (4.4). That is, the system performs as a one-fold equal-strength diversity system with an energy-to-noise ratio of $\alpha\tilde{\varepsilon}$.

The system performance depends on the filter characteristics only through the quantity $\tilde{\varepsilon}$. Since $\tilde{\varepsilon}$ cannot exceed unity and since this value is attained when the channel is nondispersive, it is natural to regard $\tilde{\varepsilon}$ as a measure of the efficiency of the system of Fig. 7.4 relative to a nondispersive system. Alternately if $u(t)$ is used as the basis for an explicit D_e-fold equal-strength diversity system and if each constituent part of the signal is processed by a demodulator of the form shown in Fig. 7.4, then the system performs as though it is an equal-strength diversity system with an energy-to-noise ratio per diversity path of $\alpha\tilde{\varepsilon}/D_e$ and with a total energy-to-noise ratio of $\alpha\tilde{\varepsilon}$. In particular if the value of D_e is adjusted so that $\alpha\tilde{\varepsilon}/D_e$ equals $\alpha_p{}^\circ$, as determined from Fig. 5.4, the system performs as though it is an optimized equal-strength diversity system with an energy-to-noise ratio per information bit of $\tilde{\varepsilon}$ times that of the given system.

Optimum Filter: Let us now consider the filter impulse response that maximizes $\tilde{\varepsilon}$ for a given complex modulation waveform $u(t)$. An explicit determination of this response is, in general, quite difficult, but a useful implicit characterization does exist. Specifically $\tilde{\varepsilon}$ is maximized by choosing $v(t)$ to be an eigenfunction of $R(t, \tau)$ associated with the dominant eigenvalue λ_0. This maximizing property of the eigenfunction is a well-known result in the theory of Hermetian integral equations [1, 2]. It is also a consequence of this theory that the resulting maximum value of $\tilde{\varepsilon}$ is equal to λ_0.

The dependence of the optimum $v(t)$ upon the scattering function is best illustrated by means of an example. To that end, let us suppose that the shape of the scattering function is doubly Gaussian, that is,

$$\sigma(r, f) = \frac{2}{S} \exp - \frac{2\pi}{S^2} \left[(rB)^2 + (fL)^2 - 2rf\sqrt{(BL)^2 - S^2} \right] \qquad (7.30a)$$

and that the complex modulation is also a Gaussian-shaped pulse, that is,

$$u(t) = \frac{2^{1/4}}{\sqrt{T}} \exp - \pi \left(\frac{t}{T} \right)^2 \left[1 + j\sqrt{(TW)^2 - 1} \right]. \qquad (7.30b)$$

This system, which is discussed in Chapter 6, yields the complex correlation function given by (6.37).

By virtue of (6.37) and (6.39), the dominant eigenvalue λ_0 is given by the expression

$$\lambda_0 = \frac{2}{1 + [1 + (BT)^2 + (LW)^2 + S^2 - 2\sqrt{(TW)^2 - 1}\sqrt{(BL)^2 - S^2}]^{1/2}} .$$

$$(7.31)$$

Moreover the derivation that yields (6.39) also demonstrates that the dominant eigenfunction is [21, 22]

$$\varphi_0(t) = \frac{2^{1/4}}{\sqrt{T_r}} \exp - \pi \left(\frac{t}{T_r}\right)^2 [1 + j\sqrt{(T_r W_r)^2 - 1}], \qquad (7.32a)$$

where

$$T_r = \frac{\sqrt{T^2 + L^2}}{[1 + (BT)^2 + (LW)^2 + S^2 - 2\sqrt{(TW)^2 - 1}\sqrt{(BL)^2 - S^2}]^{1/4}}$$

$$(7.32b)$$

and

$$W_r = \frac{\sqrt{B^2 + W^2}}{[1 + (BT)^2 + (LW)^2 + S^2 - 2\sqrt{(TW)^2 - 1}\sqrt{(BL)^2 - S^2}]^{1/4}}$$

$$(7.32c)$$

Thus $v(t)$ should be set equal to the right member of (7.32a). The resulting value of $\tilde{\varepsilon}$ then equals the right member of (7.31).

The qualities T_r and W_r are, respectively, the time duration and the bandwidth of $\varphi_0(t)$. Although the general expressions for T_r and W_r are somewhat involved, they do clearly indicate that the time duration and bandwidth of the filter impulse response may significantly exceed those of the filter matched to $u(t)$ when B or L are large. This is to be expected since the energy in the "signal" component of the received waveform is then spread over a time interval that greatly exceeds T sec or a frequency interval that greatly exceeds W Hz.

Optimum Modulation: Thus far we have maximized $\tilde{\varepsilon}$ only with respect to the demodulator filter; it is also worthwhile to maximize it with respect to the complex modulation $u(t)$. This double maximization problem possesses a strong symmetry that is best seen when (7.29) is expressed as

$$\tilde{\varepsilon} = \iint \mathscr{R}(\alpha, \beta)\theta_u(\beta, \alpha)\theta_v^*(\beta, \alpha) \, d\alpha \, d\beta, \qquad (7.33)$$

where as before

$$\mathscr{R}(\alpha, \beta) = \iint \sigma(r, f) \exp j2\pi(r\alpha + f\beta) \, dr \, df, \qquad (7.34a)$$

$$\theta_u(\beta, \alpha) = \int u\left(t - \frac{\beta}{2}\right) u^*\left(t + \frac{\beta}{2}\right) \exp j2\pi\alpha t \, dt, \qquad (7.34b)$$

and

$$\theta_v(\beta, \alpha) = \int v\left(t - \frac{\beta}{2}\right) v^*\left(t + \frac{\beta}{2}\right) \exp j2\pi\alpha t \, dt. \qquad (7.34c)$$

This symmetry, in conjunction with some other observations, suggests that the maximum value of $\tilde{\varepsilon}$ often is attained when $u(t)$ equals $v(t)$. That is, when the system of Fig. 7.4 is optimized with respect to both $u(t)$ and $v(t)$, the resulting filter is often matched to $u(t)$.

We have not ascertained the most general conditions under which the aforestated result is valid. That it is not true in general is established by the rather artificial counterexample in which the channel introduces a range delay and Doppler shift but causes no dispersion. However it is true for all those channels for which $\mathscr{R}(\alpha, \beta)$ is a real positive function.

In particular by virtue of Schwarz's inequality [23],

$$\left| \iint \mathscr{R}(\alpha, \beta) \theta_u(\beta, \alpha) \theta_v^*(\beta, \alpha) \, d\alpha \, d\beta \right|^2 \leq \left[\iint |\mathscr{R}(\alpha, \beta)| \, |\theta_u(\beta, \alpha)|^2 \, d\alpha \, d\beta \right]$$

$$\times \left[\iint |\mathscr{R}(\alpha, \beta)| \, |\theta_v(\beta, \alpha)|^2 \, d\alpha \, d\beta \right].$$

The right member of this expression attains its maximum value when $u(t)$ and $v(t)$ are identical. Consequently

$$\max \tilde{\varepsilon} \leq \max_{u(t)} \left[\iint |\mathscr{R}(\alpha, \beta)| \, |\theta_u(\beta, \alpha)|^2 \, d\alpha \, d\beta \right], \tag{7.35}$$

where the left-hand maximization is over $u(t)$ and $v(t)$. On the other hand, the value of $\tilde{\varepsilon}$ can be made equal to the right member of (7.35) when $\mathscr{R}(\alpha, \beta)$ equals $|\mathscr{R}(\alpha, \beta)|$ by setting $v(t)$ equal to $u(t)$.

We have now completed our discussion of nondispersive channels and of demodulators that tend to become optimum as the channel spread factor approaches zero. We turn next to a discussion of channels that appear to be noticeably dispersive in only one dimension. This discussion suggests demodulator structures that sometimes provide reasonable substitutes for the unwieldy systems required to implement the optimum demodulator.

7.4 SINGLY DISPERSIVE CHANNELS

We first establish the form of the correlation kernel demodulator for channels that are dispersive only in time or only in frequency. These forms may be modified to apply to some other channels by invoking the equivalent channel concepts of Chapter 2. In fact, it would suffice to restrict the present discussion to channels that are dispersive only in time. However the discussion of frequency dispersive channels provides a convenient framework for introducing some of the structures to be considered.

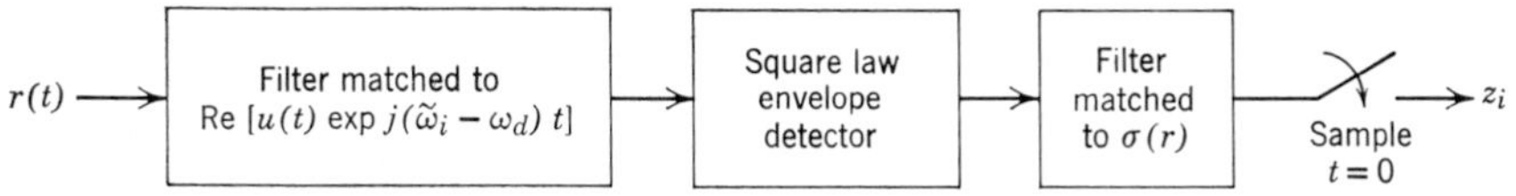

Fig 7.5 Correlation kernel demodulator for channel dispersive only in time.

Channels Dispersive Only in Time

We suppose that a correlation kernel demodulator is employed. The output of the ith branch demodulator is then

$$z_i = \iint r(t)R(t, \tau)r(\tau) \exp j\tilde{\omega}_i(t - \tau) \, dt \, d\tau. \qquad (7.36a)$$

By virtue of the assumption that the channel is only time dispersive

$$R(t, \tau) = \left[\int \sigma(\rho)u(t - \rho)u^*(\tau - \rho) \, d\rho \right] \exp j\omega_d(\tau - t), \qquad (7.36b)$$

where

$$\sigma(r, f) = \sigma(r) \, \delta(f - f_d) \qquad (7.36c)$$

and where ω_d denotes the common Doppler shift of the scatterers.

Upon introducing the right member of (7.36b) into the right member of (7.36a), we obtain

$$z_i = \int \sigma(\rho) \, d\rho \left| \int u(t - \rho)r(t) \exp j(\tilde{\omega}_i - \omega_d)t \, dt \right|^2. \qquad (7.37)$$

The right member of this equation may be interpreted as illustrated in Fig. 7.5. In the figure, the first filter is matched to the signal Re $[u(t) \exp j(\tilde{\omega}_i - \omega_d)t]$. The output of the matched filter is then passed through a square-law envelope detector and through the filter which is "matched" to the range-delay scattering function. The output of the latter filter is sampled at time t equal to zero.

The structure of Fig. 7.5 is equivalent to an optimum demodulator only if the positive eigenvalues of $R(t, \tau)$ are all equal and if in addition the channel is only time dispersive. On the other hand, the relative simplicity of the structure may warrant its use even when it is not optimal. We return to this question subsequently, but first we consider the form of the correlation kernel demodulator for channels that are dispersive only in frequency.

Channels Dispersive Only in Frequency

Channels that are only frequency dispersive possess scattering functions of the form

$$\sigma(r, f) = \delta(r - r_o)\sigma(f).$$

The corresponding complex correlation function is

$$R(t, \tau) = u(t - r_0)\left[\int \sigma(f) \exp j2\pi f(\tau - t) \, df\right] u^*(\tau - r_0). \qquad (7.38)$$

A more useful form of (7.38) is obtained by introducing the two-frequency correlation function $\mathscr{R}(\alpha, \beta)$. Specifically

$$R(t, \tau) = u(t - r_0)\mathscr{R}(0, \tau - t)u^*(\tau - r_0). \qquad (7.39)$$

By virtue of (7.36) and (7.39), the output of the ith branch of a correlation kernel demodulator for a channel dispersive only in frequency may be expressed as

$$z_i = \iint r(t)u(t - r_0)\mathscr{R}(0, \tau - t)u^*(\tau - r_0)r(\tau) \exp j\tilde{\omega}_i(t - \tau) \, dt \, d\tau. \qquad (7.40)$$

The right member of this equation may be interpreted in a variety of ways; two such interpretations are now described [4, 5, 8, 12].

Filter-Correlator: It is readily verified that $\mathscr{R}(0, \tau - t)$ possesses conjugate symmetry in the variables t and τ. Thus by the same line of reasoning that leads to the structure of Fig. 7.1, we conclude that the branch demodulator output is given by the expression

$$z_i = \text{Re}\left[\int r(\tau)u^*(\tau - r_0) \, d\tau \int r(t)u(t - r_0)g_i(\tau - t) \, dt\right], \qquad (7.41a)$$

where

$$g_i(t) = 2[\mathscr{R}(0, t) \exp -j\tilde{\omega}_i t], \qquad \text{for } t \geq 0,$$
$$g_i(t) = 0, \qquad\qquad\qquad\qquad \text{elsewhere.} \qquad (7.41b)$$

To determine the nature of (7.41a), let us first suppose that $u(t)$ is real; that is, no phase modulation is employed. The right member of (7.41) then becomes

$$z_i = \int r(\tau)u(\tau - r_0) \, d\tau \int \{\text{Re } [g_i(\tau - t)]\}r(t)u(t - r_0) \, dt. \qquad (7.42)$$

Thus as illustrated in Fig. 7.6a, z_i may be computed by first multiplying $r(t)$ by $u(t - r_0)$, passing the product through the realizable linear time-invariant filter whose impulse response is $\text{Re }[g_i(t)]$, multiplying the filter output by its input and finally integrating the resulting product.

The interpretation of (7.41a) is more complex when $u(t)$ is not real because we must then multiply the received waveform by both the real and the imaginary parts of $u(t - r_0)$ and pass the resulting products through filters whose impulse responses are the real and imaginary parts of $g_i(t)$. The resulting receiver structure is shown in Fig. 7.6b for the situation in which $g_i(t)$ is real. Although the structure of Fig. 7.6b is sometimes useful, there is an

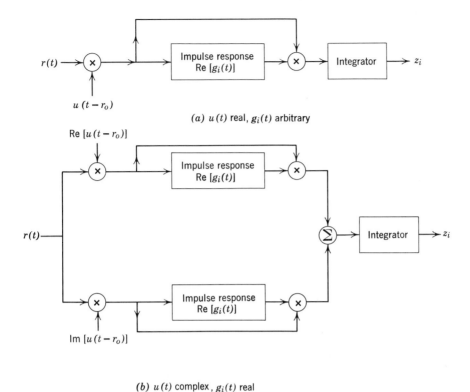

(a) $u(t)$ real, $g_i(t)$ arbitrary

(b) $u(t)$ complex, $g_i(t)$ real

Fig. 7.6 Correlation kernel demodulator for channel dispersive only in frequency.

alternate structure that is often preferable. The latter structure, which is depicted in Fig. 7.7, performs approximately the same operations as that of Fig. 7.6b. Precisely stated, it can be shown to perform the same operations if the bandwidths of the received waveform and of the filter are sufficiently less than the "carrier" frequency $\tilde{\omega}_i$ [4–6].

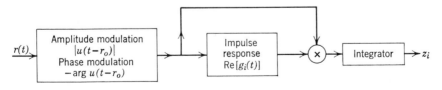

Fig. 7.7 Phase modulator implementation of correlation kernel demodulator for channels dispersive only in frequency.

Filter-Squarer-Integrator: It is sometimes desirable to avoid the use of multipliers in the branch demodulators. We now consider an implementation that does not require them.

Since $\sigma(f)$ is nonnegative and integrable, $[\sigma(f)]^{1/2}$ is square integrable and possesses a Fourier transform $h(x)$. Hence $\mathcal{R}(0, t)$ is the correlation of $h(x)$ with $h^*(x)$; that is,

$$\mathcal{R}(0, t) = \int h^*(t + x)h(x)\, dx, \tag{7.43a}$$

where

$$h(x) = \int [\sigma(f)]^{1/2} \exp -j2\pi fx\, df. \tag{7.43b}$$

We next introduce the right member of (7.43a) into (7.40) and change the variable of integration to obtain

$$z_i = \int dy \left| \int r(t)u^*(t - r_o)h_i(y - t)\, dt \right|^2, \tag{7.44a}$$

where

$$h_i(t) = h(t) \exp j\tilde{\omega}_i t. \tag{7.44b}$$

According to the comments following (7.27), the y integrand of (7.44a) is the squared envelope of the quantity

$$\mathrm{Re}\left[\int r(t)u^*(t - r_o)h_i(y - t)\, dt\right],$$

and it may be computed with a structure patterned after that of Fig. 7.2b. The output of that structure is then passed through an integrator to obtain z_i. Thus, for narrow-band signals, the complete branch demodulator can be implemented as shown in Fig. 7.8.

The filter in Fig. 7.8 is sometimes unrealizable in the sense that its impulse response may be nonzero for negative values of the argument. This is not of critical importance because the impulse response of the filter usually vanishes for sufficiently large values of either positive or negative time. Consequently the performance of the unrealizable filter may be approached as closely as desired by employing a realizable filter whose impulse response is

$$w_i(t) = \begin{cases} h_i(t - \delta), & \text{for } t \geq 0, \\ 0, & \text{for } t < 0, \end{cases} \tag{7.45}$$

where δ is chosen sufficiently large to ensure the desired degree of approximation. On the other hand, it is often difficult to implement filters that approximate unrealizable filters, and so it is of interest to determine the conditions for which the filter of Fig. 7.8 is realizable. That we now do.

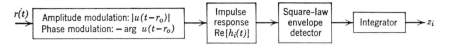

Fig. 7.8 Filter-squarer-integrator demodulator for channel dispersive only in frequency.

✍ *Realizability Conditions:* Any impulse response that satisfies (7.43a) may be used to implement the correlation kernel demodulator in the form of Fig. 7.8. Thus the question of realizability is more properly stated as: Is (7.43a) satisfied by some realizable impulse response? For the answer to the question to be in the affirmative, the Fourier transform of the two members of the equation must be identical. On the other hand, if $h(t)$ satisfies some mild mathematical constraints, the transform of the right member of (7.43) equals the squared magnitude of the inverse transform of $h(t)$ [24, 25]; that is,

$$\sigma(f) = |H(f)|^2, \tag{7.46a}$$

where

$$H(f) = \int h(t) \exp j2\pi ft \, dt. \tag{7.46b}$$

If a solution of (7.43a) is to be a realizable impulse response, its Fourier transform must satisfy (7.46). However by virtue of the Paley-Wiener theorem, the (inverse) Fourier transform of a realizable impulse response satisfies the inequality [26, 27]

$$\int \left| \frac{\ln |H(f)|^2}{1 + (2\pi f)^2} \right| df < \infty.$$

Therefore a necessary condition for a realizable impulse response to satisfy (7.43a) is that

$$\int \left| \frac{\ln \sigma(f)}{1 + (2\pi f)^2} \right| df < \infty. \tag{7.47}$$

Conversely if the last inequality is satisfied, a realizable impulse response exists whose transform satisfies (7.46a). Moreover if this impulse response, say $h(t)$, satisfies some mild mathematical constraints, the transform of $|H(f)|^2$ equals the correlation of $h(t)$ with its conjugate, and so $h(t)$ satisfies (7.43a) [24, 25].

The discussion of correlation kernel demodulators for singly dispersive channels is now complete. The relative simplicity of these structures provides a partial justification for their use, be they optimum or not. Moreover we suspect that the system of Fig. 7.5 should perform as well as a correlation kernel demodulator whenever BT is much less than unity. Similarly, the systems of Figs. 7.7 and 7.8 should perform as a correlation kernel demodulator whenever LW is much less than unity. Unfortunately neither of these comments sheds much light upon the character of "reasonable" demodulator structures for channels that are appreciably spread in both dimensions. We turn now to a discussion of that question.

7.5 DOUBLY DISPERSIVE CHANNELS

We begin by considering a restricted class of doubly dispersive channels for which the correlation kernel demodulator is relatively simple to implement. This discussion suggests an appealing means of approximating the operations performed by more general correlation kernel demodulators.

Demodulator Structure

Let us first suppose that the channel scattering function may be expressed as a weighted sum of scattering functions; that is,

$$\sigma(r, f) = \sum_i p_i \sigma_i(r, f), \tag{7.48}$$

where each of the $\sigma_i(r, f)$ is a scattering function and the p_i are positive numbers that sum to unity. When the scattering function is of this form, it is sometimes possible to decompose each of the correlation kernel branch demodulators into a collection of simpler demodulators. This is particularly true when the $\sigma_i(r, f)$ are singly dispersive.

Specifically the complex correlation function associated with (7.48) is

$$R(t, \tau) = \sum_i p_i \hat{R}_i(t, \tau), \tag{7.49}$$

where

$$\hat{R}_i(t, \tau) = \int u(t - r)\sigma_i(r, f)u^*(\tau - r) \exp j2\pi f(\tau - t) \, dr \, df. \tag{7.50}$$

Equations 7.2 and 7.49a imply that the output z_k of the kth correlation kernel branch demodulator is expressible as

$$z_k = \sum_i \hat{z}_{ik}, \tag{7.51a}$$

where

$$\hat{z}_{ik} = p_i \iint r(t)\hat{R}_i(t, \tau)r(\tau) \exp j\tilde{\omega}_k(t - \tau) \, dt \, d\tau. \tag{7.51b}$$

Thus z_k may be interpreted as the sum of the elementary outputs $\hat{z}_{ik}$.

If the constituent scattering functions $\sigma_i(r, f)$ are spread in only one dimension, the evaluation of the $\hat{z}_{ik}$ is straightforward, and the complete demodulator is relatively easy to implement—provided that the number of terms in (7.48) is not too large. Of course, as the number of terms increases, the implementation becomes impractical. However (7.51) then suggests an appealing alternate structure.

To introduce this alternate structure, we first suppose that the scattering function may be expressed as

$$\sigma(r, f) = \left[\sum_{i=1}^{N} p_i \, \delta(f - \delta_i) \right] \sigma(r), \qquad (7.52)$$

where $p_i \geq 0$ and $\sum_i p_i = 1$. That is, the scattering function consists of a sum of constituent scattering functions, each of which is spread only in range delay. Moreover $\sigma(r, f)$ is separable in that it consists of a product of a function of range delay times a function of frequency. These suppositions, in conjunction with (7.51), permit us to cast the demodulator in the form of Fig. 7.9a.

It is clear that the complexity of Fig. 7.9a increases as N increases. Since this complexity is controlled, primarily by the need for N different filters and square-law devices, a significant simplification results if, as illustrated by Fig. 7.9b, we first sum the filter outputs and then pass the resulting waveform through a single square-law envelope detector. The loss introduced by this interchange is not too pronounced if the outputs of the different filters are nearly orthogonal. Although the demodulator of Fig. 7.9b does not perform as a correlation kernel demodulator, its simplicity is so appealing that it is often preferred to the structure of Fig. 7.9a. This is particularly true when N is large since the structure of Fig. 7.9a becomes more and more complicated with increasing N, whereas increasing N does not appreciably complicate the structure of Fig. 7.9b. In fact if we let N increase in such a way that $\sum_i \sqrt{p_i} \exp -j2\pi\delta_i t$ approaches a limit $w(t)$, the structure of Fig. 7.9b reduces to the simpler form shown in Fig. 7.10.

Although the demodulator of Fig. 7.10 results from rather arbitrary simplifications of structures that are optimum under restrictive conditions, the correlation kernel demodulator sometimes may be implemented precisely in the form of Fig. 7.10. This is true, for example, if the scattering function is doubly Gaussian shaped and if the complex modulation is also Gaussian shaped, that is, if they are of the form described by (7.30).

Outline of Performance Analysis

Although the demodulator structure of Fig. 7.10 has much to recommend it, its performance is usually inferior to both that of the optimum demodulator

206

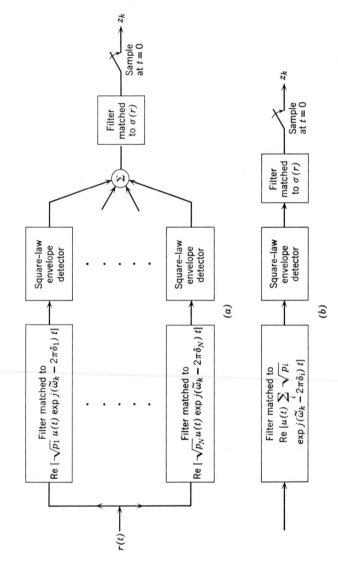

Fig. 7.9 Demodulators for a doubly dispersive channel; (*a*) Correlation kernel demodulator; (*b*) Suboptimum demodulator.

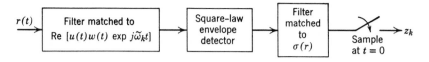

Fig. 7.10 Demodulator of Fig. 7.9b in the limit as $\sum_i \sqrt{p_i} \exp -j2\pi\delta_i\, t$ approaches $\omega(t)$.

and that of the correlation kernel demodulator. However the derivation of these structures does suggest that they should perform well when the positive eigenvalues are all equal and when in addition the channel appears to be spread in a single dimension. Quantitative measures of performance can be obtained from the analysis procedure outlined below.

The analysis of systems employing the branch demodulators of Fig. 7.10 is easily reduced to the determination of the joint probability density of the m branch demodulator outputs, given that a particular waveform is transmitted. This determination is greatly simplified when the differences between the $\tilde{\omega}_i$ are so large that the outputs of the different branches are statistically independent of each other for the stated conditioning. Then we need only determine the statistical properties of each branch demodulator output considered separately. This supposition is introduced in the discussion of Chapter 4 (Fig. 4.9) and again in the discussion of the filter-squarer demodulator for slightly dispersive channels. We suppose it is valid in the subsequent discussion.

The statistical properties we seek can be ascertained rather easily once Fig. 7.10 is recast in the form of Fig. 7.11. This is true because the waveform $y(t)$ in Fig. 7.11 is a linear transformation of the received process $r(t)$ and because $r(t)$ is a zero-mean Gaussian random process. Consequently $y(t)$ is also a zero-mean Gaussian random process, and the required statistical properties can be obtained by straightforward but tedious calculation.

Specifically $y(t)$ can be expanded in a Karhunen-Loève series with statistically independent zero-mean Gaussian random coefficients, and the outputs of the branch demodulators may then be expressed as the sum of the squares of these coefficients. Such sums were encountered in the analysis of optimum

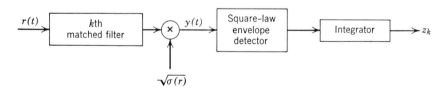

Fig. 7.11 Representation of demodulator output as the integrated square of a Gaussian random process.

demodulators and their moment-generating functions were determined. The corresponding moment-generating functions for the system of Fig. 7.11 can be obtained in a similar manner. Once they are obtained, we appeal to the error-probability bounds of (4.30) and the probability distribution bounds of Appendix 3 to obtain bounds to the system error probability. Numerical techniques are usually required to evaluate these bounds.

The discussion of demodulator structures is now complete. Our principal conclusions are that the optimum demodulator is frequently very difficult to implement but that there are conditions for which simpler structures can be employed with little loss in performance. Although the simplest of all structures is clearly the filter-squarer system, its performance deteriorates severely as the dominant eigenvalue diminishes from a value of unity, that is, as the number of diversity paths increases from one. Other structures, such as the filter-squarer-filter, sometimes perform quite well when the dominant eigenvalue is less than unity. This is particularly true when the channel is only dispersive in time. When a quantitative measure of the loss is required, it can be determined by completing the analysis outlined above.

REFERENCES

[1] R. Courant and D. Hilbert, *Methods of Mathematical Physics*, Vol. 1. New York: Interscience, 1953, Chapter 3.

[2] F. G. Tricomi, *Integral Equations*. London: Interscience, 1957. Chapters 2 and 3.

[3] F. Smithies, *Integral Equations*. London: Cambridge University Press, 1958.

[4] J. V. Evans and T. Hagfors, Ed, *Radar Astronomy*. R. Price, New York: McGraw-Hill, 1968, Chapter 10, p. 547.

[5] J. C. Hancock and P. A. Wintz, *Signal Detection Theory*. New York: McGraw-Hill, 1966, Chapter 7.

[6] E. J. Baghdady, Ed, *Lectures on Communication System Theory*. J. M. Wozencraft, New York: McGraw-Hill, 1961, Chapter 12, p. 279.

[7] C. W. Helstrom, *Statistical Theory of Signal Detection*. New York: Pergamon, 1960, Chapter 11.

[8] H. L. Van Trees, *Detection, Estimation, and Modulation Theory*. part 2, New York: Wiley, 1969, Chapter 3.

[9] R. Price, "Optimum Detection of Random Signals in Noise, with Application to Scatter-Multipath Communication, I." *IRE Trans. Inform. Theory*, pp. 125–135, December 1956.

[10] G. L. Turin, "Communication through Noisy, Random-Multipath Channels." *IRE Convention*, pp. 154–166, 1956.

[11] D. Middleton, "On the Detection of Stochastic Signals in Additive Normal Noise— Part I." *IRE Trans. Inform. Theory*, pp. 86–121, June 1957.

[12] T. Kailath, "Correlation Detection of Signals Perturbed by a Random Channel." *IRE Trans. Inform. Theory*, pp. 361–366, June 1960.

[13] D. J. Sakrison, *Communication Theory: Transmission of Waveforms and Digital Information*. New York: Wiley, 1968, Chapter 5.

[14] C. W. Helstrom, *Statistical Theory of Signal Detection*. New York: Pergamon, 1960, Chapter 1.

[15] M. Schwartz, W. R. Bennett, and S. Stein, *Communication Systems and Techniques*. New York: McGraw-Hill, 1966, Chapter 1.

[16] J. J. Downing, *Modulation Systems and Noise*. New Jersey: Prentice-Hall, Chapter 3, 1964.

[17] J. M. Wozencraft and I. M. Jacobs, *Principles of Communication Engineering*. New York: Wiley, 1965, pp. 233–245.

[18] D. J. Sakrison, *Communication Theory: Transmission of Waveforms and Digital Information*. New York: Wiley, 1968, pp. 200 and 240.

[19] W. D. Davenport, Jr., and W. L. Root, *Random Signals and Noise*. New York: McGraw-Hill, 1958, pp. 158–165.

[20] J. M. Wozencraft and I. M. Jacobs, *Principles of Communication Engineering*. New York: Wiley, 1965, p. 114.

[21] R. Price and P. Green, "Signal Processing in Radar Astronomy-Communication via Fluctuating Multipath Media," Lincoln Laboratory, MIT, Tech. Rept. 234, pp. C78–C85, DDC No. 246782, 1960.

[22] N. Wiener, *The Fourier Integral and Certain of Its Applications*. New York: Dover, 1932, pp. 55–64.

[23] G. H. Hardy, J. E. Littlewood, and G. Polya, *Inequalities*. London: Cambridge University Press, 1952, p. 132.

[24] D. J. Sakrison, *Communication Theory: Transmission of Waveforms and Digital Information*. New York: Wiley, 1968, pp. 30 and 31.

[25] A. A. Papoulis, *The Fourier Integral and Its Applications*. New York: McGraw-Hill, 1962, Chapter 2.

[26] N. Wiener, *The Extrapolation, Interpolation, and Smoothing of Stationary Time Series*. New York: Wiley, 1949.

[27] E. A. Guillemin, *Theory of Linear Physical Systems*. New York: Wiley, 1963.

Appendix 1

Central Limit Theorem for the Point-Scatterer Model

In this appendix we determine conditions that imply that any given subset of the Karhunen-Loève coefficients z_k converges, in distribution, to statistically independent Gaussian random variables.

The variates in question are determined by the expressions

$$z_k = \left(\frac{2}{E_r}\right)^{1/2} \int y(t)\varphi_k^*(t) \exp - j\omega_o t \, dt, \qquad (A1.1)$$

where the $\varphi_i(t)$ are the eigenfunctions of the complex correlation function $R(t, \tau)$ (Fig. 2.2). The joint characteristic function of any subset of N such complex variates is defined as

$$M(\mathbf{v}) = \overline{\exp j \frac{1}{2} \sum_{k=1}^{N} (z_k v_k^* + z_k^* v_k)}, \qquad (A1.2)$$

where $\mathbf{v}$ is an N-dimensional vector with complex elements and the bar denotes an ensemble average. If the variates are statistically independent zero-mean Gaussian random variables with $\overline{z_i^2} = 0$, this characteristic function becomes

$$M(\mathbf{v}) = \exp -\frac{1}{4} \sum_{k=1}^{N} \lambda_k |v_k|^2, \qquad (A1.3)$$

where λ_i is the eigenvalue associated with $\varphi_i(t)$.

We now show that the right member of (A1.2) converges to the right member of (A1.3) if all of the fractional scatterer cross sections tend to zero. By the properties of characteristic functions [1], this in turn ensures that the z_i converge to zero-mean statistically independent Gaussian random variables with variances λ_i.

As a first step in the derivation, we recall from (2.8) that the received waveform may be expressed as

$$y(t) = A \operatorname{Re} \left(\sum_i \rho_i u(t - r_i) \exp j[(\omega_o - \omega_i)t - \omega_o r_i - \theta_i] \right) \quad \text{(A1.4)}$$

Upon introducing the right member of (A1.4) into (A1.1), we obtain

$$z_k = \left(\sum_i \rho_i B_{ik} \exp - j\theta_i \right) + \left(\sum_i \rho_i \tilde{B}_{ik} \exp j\theta_i \right), \quad \text{(A1.5)}$$

where

$$B_{ik} = \frac{A}{\sqrt{2E_r}} (\exp - j\omega_o r_i) \int u(t - r_i)\varphi_k^*(t) \exp - j\omega_i t \, dt \quad \text{(A1.6}a)$$

and

$$\tilde{B}_{ik} = \frac{A}{\sqrt{2E_r}} (\exp j\omega_o r_i) \int u^*(t - r_i)\varphi_k^*(t) \exp - j(2\omega_o - \omega_i)t \, dt. \quad \text{(A1.6}b)$$

We next impose the narrow-band assumption of Chapter 2; that is, we suppose that the right member of (A1.6b) vanishes for all values of i and k. This is almost equivalent to the supposition (2.47) that

$$\int \varphi_k(t)\varphi_i(t) \exp j2\omega_o t \, dt = 0.$$

The present supposition is approximately satisfied for any given system if ω_o is sufficiently large. Alternately the right member of (A1.6b) is identically zero if $u(t)$ and $\varphi_i(t)$ are strictly bandlimited to the band $-\Omega/2$ to $+\Omega/2$ rad/sec and if in addition $2\omega_o$ exceeds the sum of $\Omega/2$ and the maximum value of $|\omega_i|$. The justification for this assertion is similar to that employed in conjunction with (2.47); hence it is not pursued here.

By virtue of our supposition, (A1.2) may be restated as

$$M(\mathbf{v}) = \overline{\exp j \left[\sum_i \rho_i B_i \cos (\varphi_i - \theta_i) \right]}, \quad \text{(A1.7)}$$

where

$$B_i \exp j\varphi_i = \sum_k B_{ik} v_k^* \quad \text{(A1.8)}$$

and where B_i is taken to be real and positive. Since the θ_i are statistically independent of each other, $M(\mathbf{v})$ may also be expressed as

$$M(\mathbf{v}) = \prod_i \overline{\exp [j\rho_i B_i \cos (\varphi_i - \theta_i)]} \quad \text{(A1.9)}$$

for any fixed given values of ρ_i and B_i.

We next recall that each of the θ_i are uniformly distributed in the interval $-\pi$ to π. Consequently the averages appearing in (A1.9) are given by the expression.

$$\overline{\exp\left[j\rho_i B_i \cos(\varphi_i - \theta_i)\right]} = \frac{1}{2\pi} \int_{-\pi}^{\pi} \exp\left[j\rho_i B_i \cos(\varphi_i - \theta)\right] d\theta. \quad \text{(A1.10)}$$

The right member of this expression may be evaluated to obtain $J_o(\rho_i B_i)$, the zero-order Bessel function of the first kind [2]. Thus we obtain

$$M(\mathbf{v}) = \prod_i J_o(\rho_i B_i). \quad \text{(A1.11)}$$

The value of $J_o(x)$ is positive for all values of x in the closed interval $[0, 1]$. Consequently $M(\mathbf{v})$ is a real positive function for all values of the v_k such that

$$\rho_i B_i \leq 1. \quad \text{(A1.12)}$$

Moreover by expanding $J_o(x)$ in a Taylor series with a remainder, we can show that

$$1 - \left(\frac{\rho_i B_i}{2}\right)^2 \leq J_o(\rho_i B_i) \leq 1 - \left(\frac{\rho_i B_i}{2}\right)^2 + d\left(\frac{\rho_i B_i}{2}\right)^4$$

provided that $\rho_i B_i$ does not exceed 1, where d is a constant independent of $\rho_i B_i$ [3]. Therefore subject to (A1.12),

$$\prod_i \left[1 - \left(\frac{\rho_i B_i}{2}\right)^2\right] \leq M(\mathbf{v}) \leq \prod_i \left[1 - \left(\frac{\rho_i B_i}{2}\right)^2 + d\left(\frac{\rho_i B_i}{2}\right)^4\right] \quad \text{(A1.13)}$$

We next observe that, for any set of positive numbers $\{x_i\}$ that do not exceed unity,

$$\prod_i (1 - x_i) = \exp \sum_i \ln(1 - x_i) \geq \exp - \left(1 + \frac{a}{1-a}\right) \sum_i x_i,$$

where a is any upper bound to the x_i. Applying this inequality to the leftmost member of (A1.13) yields

$$\prod_i \left[1 - \left(\frac{\rho_i B_i}{2}\right)^2\right] \geq \exp - \left(1 + \frac{c}{1-c}\right) \sum_i \left(\frac{\rho_i B_i}{2}\right)^2, \quad \text{(A1.14)}$$

where c is any upper bound to the $(\rho_i B_i/2)^2$. By a similar argument

$$\prod_i \left[1 - \left(\frac{\rho_i B_i}{2}\right)^2 + d\left(\frac{\rho_i B_i}{2}\right)^4\right] \leq \exp(-1 + dc) \sum_i \left(\frac{\rho_i B_i}{2}\right)^2. \quad \text{(A1.15)}$$

To proceed, we evaluate the summation appearing in (A1.14) and (A1.15). By virtue of (A1.8)

$$\sum_i (\rho_i B_i)^2 = \sum_k \sum_n v_k^* v_n \sum_i \rho_i^2 B_{ik} B_{in}^*, \tag{A1.16}$$

or upon combining (A1.6a) and (A1.16),

$$\sum_i (\rho_i B_i)^2 =$$

$$\sum_k \sum_n \frac{v_k^* v_n}{E_r} \iint \varphi_k^*(t) \varphi_n(\tau) \frac{A^2}{2} \rho_i^2 u(t - r_i) u^*(\tau - r_i) \exp - j\omega_i(t - \tau) \, dt \, d\tau.$$

Finally by comparison with (2.10), (2.11), and (2.16), we find that the summation over i is proportional to the complex correlation function $R(t, \tau)$ associated with $u(t)$. Specifically

$$\sum_i (\rho_i B_i)^2 = \sum_k \sum_n v_k^* v_n \iint \varphi_k^*(t) R(t, \tau) \varphi_n(\tau) \, dt \, d\tau,$$

or since the $\varphi_k(t)$ are normalized eigenfunctions of $R(t, \tau)$,

$$\sum_i (\rho_i B_i)^2 = \sum_k |v_k|^2 \lambda_k, \tag{A1.17}$$

where the λ_k are the eigenvalues of $R(t, \tau)$.

A combination of (A1.13) to (A1.15), and (A1.17) yields

$$\exp - \frac{1}{4} \left(\frac{c}{1 - c} \right) \sum_k \lambda_k |v_k|^2 \leq M(\mathbf{v}) \left\{ \exp - \frac{1}{4} \sum_{k=1}^N \lambda_k |v_k|^2 \right\}^{-1} \leq \exp \frac{dc}{4} \sum_k \lambda_k |v_k|^2 \tag{A1.18}$$

provided that

$$\left| \frac{\rho_i B_i}{2} \right|^2 \leq c \leq \frac{1}{4} \tag{A1.19}$$

for all i.

It is clear that the rightmost and leftmost members of (A1.18) tend to unity as c tends to zero. This implies that $M(\mathbf{v})$ converges to the characteristic function of a set of statistically independent zero-mean complex Gaussian random variables with variances λ_k. Thus the conditions we now seek are those which imply the vanishing of c for any fixed values of the (complex) v_k, $k = 1, \ldots, N$.

By virtue of (A1.6a) and (A1.8), the value of $(\rho_i B_i/2)^2$ is given by the expression

$$\left(\frac{\rho_i B_i}{2} \right)^2 = A^2 \frac{(\rho_i)^2}{8E_r} \left| \sum_k v_k^* \int u(t - r_i) \varphi_k^*(t) \exp - j\omega_i t \, dt \right|^2.$$

We next overbound the sum to obtain

$$\left(\frac{\rho_i B_i}{2}\right)^2 \leq \frac{A^2(\rho_i)^2}{8E_r} \left\{\sum_k |v_k| \left| \int u(t - r_i)\varphi_k^*(t) \exp - j\omega_i t \, dt \right|\right\}^2,$$

or by virtue of Schwarz's inequality,

$$\left(\frac{\rho_i B_i}{2}\right)^2 \leq \frac{A^2(\rho_i)^2}{8E_r} \int |u(t)|^2 \, dt \left(\sum_{k=1}^{N} |v_k|\right)^2.$$

On the other hand, it follows from (2.8), (2.13), and (2.15) that

$$E_r = \frac{A^2}{2}\left(\sum_n |\rho_n|^2\right) \int |u(t)|^2 \, dt.$$

Hence

$$\left(\frac{\rho_i B_i}{2}\right)^2 \leq \frac{0.25|\rho_i|^2}{\sum_n |\rho_n|^2}\left(\sum_{k=1}^{N} |v_k|\right)^2.$$

Consequently we may define c by the expression

$$c = \max_i \left[\frac{0.25|\rho_i|^2}{\sum_n |\rho_n|^2}\left(\sum_{k=1}^{N} |v_k|\right)^2\right].$$

Clearly c equals zero for any fixed values of the v_k if the fractions $|\rho_i|^2(\sum_n |\rho_n|^2)^{-1}$ are all equal to zero. Therefore if the fractional cross section of each scatterer tends to zero, then c tends to zero, the characteristic function of the variates tends to the right member of (A1.3), and the variates themselves converge in distribution to statistically independent zero-mean Gaussian random variables with variances λ_i. Thus the proof is complete.

REFERENCES

[1] W. Feller, *An Introduction to Probability Theory and Its Applications*, Vol. II. New York: Wiley, 1966, p. 481.
[2] W. Magnus and F. Oberhettinger, *Formulas and Theorems for the Functions of Mathematical Physics*. New York: Chelsea, 1954, p. 26.
[3] I. S. Sokolnikoff and R. M. Redheffer, *Mathematics of Physics and Modern Engineering*. New York: McGraw-Hill, 1958, pp. 143–144.

Derivation of Error-Probability Bounds

The mathematical results of Chapters 4 and 5 are obtained in this appendix. The functional bounds of (4.30) are derived and applied to the additive white Gaussian noise channel to obtain the results of Section 5.1. The error-probability bounds for fading dispersive channels are then derived.

A. FUNCTIONAL BOUNDS

We first seek bounds to the probability of (4.29), that is,

$$P(\varepsilon) = 1 - P(x > y_j, j = 1, \ldots, m - 1). \tag{A2.1}$$

This expression may be restated as

$$P(\varepsilon) = 1 - \int p(x)[P(y < x)]^{m-1} \, dx, \tag{A2.2}$$

where y now denotes the random output of any one of the incorrect branch demodulators. The validity of (A2.2) hinges upon the fact that the y_i are statistically independent of each other and are identically distributed random variables.

1. Upper Bound

We next observe that

$$
\int p(x)[P(y < x)]^{m-1} \, dx = \int_0^h p(x)[P(y < x)]^{m-1} \, dx
$$

$$
+ \int_h^\infty p(x)[P(y < x)]^{m-1} \, dx
$$

$$
\geq \int_h^\infty p(x)[P(y < x)]^{m-1} \, dx
$$

$$
\geq \int_h^\infty p(x)[1 - (m-1)P(y \geq x)] \, dx,
$$

where we are free to choose the value of h.

Hence

$$P(\varepsilon) \leq 1 - P(x > h) + (m - 1) \int_h^\infty p(x)P(y \geq x)\,dx, \qquad \text{(A2.3)}$$

or

$$P(\varepsilon) \leq P(x \leq h) + mP(h < x \leq y). \qquad \text{(A2.4)}$$

Thus the upper bound of (4.30*a*) is established. This result has appeared previously [1].

2. Lower Bound

To obtain a suitable lower bound to the right member of (A2.2), we observe that

$$\int p(x)[P(y < x)]^{m-1}\,dx = \int p(x) \exp \left\{ (m - 1) \ln\left[1 - P(y \geq x)\right] \right\}\,dx. \qquad \text{(A2.5)}$$

But the logarithmic function satisfies the inequality

$$\ln x \leq x - 1,$$

thus the right member of (A2.5) does not exceed the quantity

$$\int p(x) \exp -(m - 1)P(y \geq x)\,dx.$$

Upon introducing this upper bound into the right member of (A2.2), we obtain

$$P(\varepsilon) \geq 1 - \int p(x) \exp -(m - 1)P(y \geq x)\,dx, \qquad \text{(A2.6)}$$

or equivalently

$$P(\varepsilon) \geq \int p(x)[1 - \exp -(m - 1)P(y \geq x)]\,dx. \qquad \text{(A2.7)}$$

The integrand of (A2.7) is nonnegative for all values of x, and so the integral can only be reduced if the integration is restricted to values of x that are less than a number h. The integral is further diminished if $P(y \geq x)$ is replaced by $P(y \geq h)$. Therefore $P(\varepsilon)$ satisfies the inequality

$$P(\varepsilon) \geq \int_\infty^h p(x)[1 - \exp - (m - 1)P(y \geq h)]\,dx, \qquad \text{(A2.8)}$$

or equivalently

$$P(\varepsilon) \geq P(x \leq h)[1 - \exp - (m - 1)P(y \geq h)]. \qquad \text{(A2.9)}$$

The bound of (A2.9) may be further simplified by noting that, for $u \geq 0$,

$$\exp - u \leq 1 - u + \frac{u^2}{2}.$$

Hence

$$P(\varepsilon) \geq \left[1 - \left(1 - \frac{1}{m} \right) \frac{m}{2} P(y \geq h) \right] \left(1 - \frac{1}{m} \right) m P(x \leq h) P(y \geq h). \quad (A2.10)$$

Finally, we upper and lower bound $(1 - 1/m)$ by 1 and $\frac{1}{2}$ (for $m \geq 2$) and suppose that the parameter h is chosen so that

$$mP(y \geq h) \leq 1. \quad (A2.11a)$$

The introduction of these bounds into (A2.10) then yields

$$P(\varepsilon) \geq \frac{m}{4} P(x \leq h) P(y \geq h) \quad (A2.11b)$$

provided that (A2.11a) is satisfied. Thus, the lower bound of (4.30b) is established.

B. THE ADDITIVE WHITE GAUSSIAN NOISE CHANNEL

The bounds of (A2.4) and (A2.11) are now applied to that channel wherein the only disturbance is additive white Gaussian noise with power density N_o watts/Hz (unilateral spectrum). Subject to the conditions discussed in Section 5.1, the random variables x and y appearing in (A2.4) and (A2.11) are independent and Gaussian with unit variances. The mean of y is zero whereas that of x is $\sqrt{2\alpha}$, where α is the energy–to–noise ratio. Consequently the probabilities $P(x \leq h)$ and $P(y \geq h)$ may be expressed as

$$P(x \leq h) = \Phi(h - \sqrt{2\alpha}) \quad (A2.12a)$$

and

$$P(y \geq h) = \Phi(-h), \quad (A2.12b)$$

where $\Phi(\cdot)$ is the probability distribution function of a zero-mean unit-variance Gaussian random variable.

1. Lower Bound

An explicit lower bound to the error probability for the additive white Gaussian noise channel is now obtained. The derivation requires the use of the following bounds to the Gaussian distribution function for negative values of the argument [2]:

$$B_1 \exp -\frac{z^2}{2} \leq \Phi(z) \leq B_2 \exp -\frac{z^2}{2}, \qquad (A2.13a)$$

where

$$B_1 = \frac{1}{\sqrt{2\pi|z|}} \left(1 - \frac{1}{z^2}\right) \qquad (A2.13b)$$

and either B_2 may be taken to be unity or

$$B_2 = \frac{1}{\sqrt{2\pi|z|}}. \qquad (A2.13c)$$

Upon combining (A2.12) and (A2.13) with (A2.11), we obtain

$$P(\varepsilon) \geq \hat{K}_1 m \exp -\tfrac{1}{2}\left[h^2 + \left(h - \sqrt{2\alpha}\right)^2\right] \qquad (A2.14a)$$

provided that

$$m\Phi(-h) \leq 1 \qquad (A2.14b)$$

and that

$$0 \leq h \leq \sqrt{2\alpha}, \qquad (A2.14c)$$

where

$$\hat{K}_1 = \frac{1}{8\pi h\left(\sqrt{2\alpha} - h\right)}\left(1 - \frac{1}{h^2}\right)\left[1 - \frac{1}{\left(h - \sqrt{2\alpha}\right)^2}\right]. \qquad (A2.14d)$$

It is clear that the inequality of (A2.14b) is satisfied if it is satisfied with $\Phi(-h)$ replaced by its upper bound, that is, if

$$m \exp -\frac{h^2}{2} \leq 1, \qquad (A2.15)$$

where the value of B_2 in the bound is, for simplicity, taken to be unity. In particular if m is determined parametrically in terms of h by the expression

$$m = \exp \tfrac{1}{2}h^2, \qquad (A2.16a)$$

the bound of (A2.14) becomes

$$P(\varepsilon) \geq \hat{K}_1 \exp -\tfrac{1}{2}\left(h - \sqrt{2\alpha}\right)^2. \qquad (A2.16b)$$

We next introduce R, the information rate of the system, and C, the channel capacity, defined as

$$R = \frac{\log_2 m}{\tau} \qquad \text{bits/sec} \qquad (A2.17a)$$

and

$$C = \frac{\alpha}{\tau} \frac{1}{\ln 2} \qquad \text{bits/sec,} \qquad (A2.17b)$$

where τ is the time allotted to the transmission of a waveform. Upon eliminating α and h from (A2.14d) and (A2.16) in favor of R, C, and τ, we obtain

$$P(\varepsilon) \geq \hat{K}_1 \cdot 2^{-\tau C \hat{E}}, \qquad (A2.18a)$$

where

$$\hat{E} = \left[1 - \left(\frac{R}{C} \right)^{1/2} \right]^2, \qquad \text{for} \quad 0 \leq \frac{R}{C} \leq 1, \qquad (A2.18b)$$

$$\hat{K}_1 = \frac{1}{16\pi\alpha\sqrt{R/C}(1 - \sqrt{R/C})} \left[1 - \frac{1}{2\alpha} \left(\frac{C}{R} \right) \right] \left[1 - \frac{1}{2\alpha(1 - \sqrt{R/C})^2} \right]$$

$$(A2.18c)$$

and α is, as before, the energy–to–noise ratio. The constraint on R/C is imposed in order to satisfy (A2.14c).

It turns out that the bound of (A2.18) is relatively sharp for values of R/C exceeding $\frac{1}{4}$, but for smaller values of R a sharper result can be obtained. In particular if h is set equal to $\sqrt{\alpha/2}$, (A2.15) is satisfied for all values of m that do not exceed $\exp \frac{1}{4}\alpha$ and (A2.14) becomes

$$P(\varepsilon) \geq K_1' m \exp - \tfrac{1}{2}\alpha$$

provided that

$$m \leq \exp \tfrac{1}{4}\alpha,$$

or in terms of R, C, and τ,

$$P(\varepsilon) \geq K_1' \cdot 2^{-\tau C(1/2 - R/C)} \qquad (A2.19a)$$

provided that R/C does not exceed $\frac{1}{4}$. The coefficient K_1' in this expression is given by (A2.18c) with R/C equal to $\frac{1}{4}$; that is,

$$K_1' = \frac{1}{4\pi\alpha} \left(1 - \frac{2}{\alpha} \right)^2. \qquad (A2.19b)$$

A comparison of (A2.18) with (A2.19) indicates that the latter bound is the sharper of the two for values of R/C less than $\frac{1}{4}$. Thus the best result is obtained by the following combination of those two equations

$$P(\varepsilon) \geq K_1 \cdot 2^{-\tau CE}, \qquad (A2.20a)$$

where

$$E = \begin{cases} \left| \dfrac{1}{2} - \dfrac{R}{C} \right|, & \text{for} \quad 0 \leq \dfrac{R}{C} \leq \dfrac{1}{4}, \\[3mm] \left[1 - \left(\dfrac{R}{C} \right)^{1/2} \right]^2, & \text{for} \quad \dfrac{1}{4} \leq \dfrac{R}{C} \leq 1, \end{cases} \qquad (A2.20b)$$

and

$$K_1 = \begin{cases} \left| \dfrac{1}{4\pi\alpha} \left(1 - \dfrac{2}{\alpha} \right)^2 \right|, & \text{for} \quad 0 \leq \dfrac{R}{C} \leq \dfrac{1}{4}, \\[3mm] \left| \dfrac{\sqrt{C/R}}{16\pi\alpha(1 - \sqrt{R/C})} \left[1 - \dfrac{1}{2\alpha}\left(\dfrac{C}{R} \right) \right]\left[1 - \dfrac{1}{2\alpha(1 - \sqrt{R/C})^2} \right], & \\[3mm] & \text{for} \quad \dfrac{1}{4} \leq \dfrac{R}{C} \leq 1. \end{cases} \qquad (A2.20c)$$

The derivation of the lower bound to the error probability is now complete, and we turn to the companion upper bound.

2. Upper Bound

The bound follows from (A2.4) upon overestimating the probabilities appearing in it. A suitable estimate for $P(x \leq h)$ is provided by (A2.12), and so we need only consider the probability that x is larger than h but no larger than y.

To upper bound this probability, we first express it as

$$P(h < x \leq y) = \int_h^\infty p(x)P(y \geq x)\, dx.$$

We next upper bound the probability $P(y \geq x)$ by an application of (A2.12b) and (A2.13) with B_2 equal to $1/\sqrt{2\pi}\,|x|$. The result is

$$P(h < x \leq y) \leq \frac{1}{\sqrt{2\pi}} \int_h^\infty \frac{1}{|x|} p(x) \exp\left(-\frac{x^2}{2}\right) dx \qquad (A2.21)$$

provided that h exceeds zero.

The right member of (A2.21) also can be expressed in the form

$$P(h < x \le y) \le \int_{h}^{\infty} \frac{1}{|x| 2\pi} \exp - \tfrac{1}{2}[(x - \sqrt{2\alpha})^2 + x^2] \, dx \quad (A2.22)$$

because $p(x)$ is the density of a Gaussian random variable with mean $\sqrt{2\alpha}$ and unit variance. Upon completing the square on the exponent and noting that $|x|$ exceeds h over the range of integration, we may reduce (A2.22) to

$$P(h < x \le y) \le \frac{1}{2h\sqrt{\pi}} \left(\exp - \frac{\alpha}{2} \right) \Phi(\sqrt{\alpha} - \sqrt{2}h), \quad (A2.23)$$

or, upon an application of (A2.13),

$$P(h < x \le y) \le \frac{1}{4\pi h \, |h - \sqrt{\alpha/2}|} \exp - \tfrac{1}{2}[\alpha + (\sqrt{\alpha} - \sqrt{2}h)^2] \quad (A2.24a)$$

provided that

$$h \ge \sqrt{\frac{\alpha}{2}}. \quad (A2.24b)$$

To proceed, (A2.12), (A2.13), and (A2.24) are combined with (A2.4) to obtain

$$P(\varepsilon) \le \hat{K}_2 \exp -\tfrac{1}{2}(h - \sqrt{2\alpha})^2 + K_2'm \exp -\tfrac{1}{2}[\alpha + (\sqrt{\alpha} - \sqrt{2}h)^2] \quad (A2.25a)$$

provided that $\sqrt{\alpha/2} \le h \le \sqrt{2\alpha}$, where

$$\hat{K}_2 = \frac{1}{\sqrt{2\pi} |h - \sqrt{2\alpha}|}, \quad (A2.25b)$$

$$K_2' = \frac{1}{4\pi h \, |h - \sqrt{\alpha/2}|}. \quad (A2.25c)$$

The derivation is completed by choosing that value of h in the interval $\sqrt{\alpha/2}$ to $\sqrt{2\alpha}$ which essentially minimizes the right member of (A2.25a). It turns out that h should be chosen to satisfy (A2.16) if R/C exceeds $\tfrac{1}{4}$ and should be set equal to $\sqrt{\alpha/2}$ otherwise. With these choices of h, the right member of (A2.25a) becomes

$$P(\varepsilon) \le K_2 \cdot 2^{-\tau CE}, \quad (A2.26a)$$

where E is defined by (A2.20b) and where

$$K_2 = \frac{1}{2\sqrt{\pi\alpha}} \left[\frac{1}{1 - \sqrt{R/C}} + \frac{\sqrt{C/R}}{2\sqrt{\pi\alpha}(2\sqrt{R/C} - 1)} \right]. \quad (A2.26b)$$

For R/C less than $\frac{1}{4}$ a better value of K_2 is obtained by setting h equal to minus infinity in (A2.4) and applying the results of (A2.12) and (A2.13) to the Gaussian variate $x - y$. The result is

$$K_2 = \frac{1}{\sqrt{2\pi\alpha}}, \qquad 0 \le \frac{R}{C} \le \frac{1}{4}. \qquad (A2.26c)$$

The results of (A2.20) and (A2.26) constitute the basic bounds to the error probability of the Gaussian channel. The upper bound derivation is, in essence, due to Fano [1, 3]. The lower bound derivation is somewhat novel, although the exponential result has been obtained previously [4].

The transition from the original bounds of (A2.4) and (A2.11) to the simpler results of (A2.20) and (A2.26) draws heavily upon the availability of sharp estimates for the Gaussian probability distribution function. In order to obtain comparable results for fading dispersive channels, the corresponding bounds to the probability distribution functions must be derived. That derivation is presented in Appendix 3. It consists of two parts. The first is devoted to a general formulation of the problem in terms of tilted probability densities and semi-invariant moment-generating functions. The second part is devoted to the specific computations for the problems of interest in Chapter 5. Since the derivation is rather lengthy, it is divorced from the central derivation of error-probability bounds. The reader may either pause at this juncture and acquaint himself with the results of Appendix 3 or proceed directly to their application in the next section.

C. FADING DISPERSIVE CHANNELS

According to (A2.4) and (A2.11), the error probability $P(\varepsilon)$ of the system satisfies the inequality

$$P(\varepsilon) \le P(x \le h) + mP(h < x \le y) \qquad (A2.27)$$

for all values of h, and it satisfies the inequality

$$P(\varepsilon) \ge \frac{m}{4} P(x \le h)P(y \ge h) \qquad (A2.28a)$$

provided that

$$mP(y \ge h) \le 1. \qquad (A2.28b)$$

In this section the lower, and then the upper, of these bounds are reduced to the results of Chapter 5.

As a preliminary remark, we recall that the variables x and y are, respectively, the random outputs of the correct and incorrect branch demodulators. By virtue of (4.26) and (A3.1), the semi-invariant, or logarithmic, moment-generating functions of the variables are, respectively,

$$\gamma_0(s) = - \sum_i \ln (1 - \alpha \lambda_i s) \tag{A2.29a}$$

and

$$\gamma_1(s) = - \sum_i \ln \left(1 - \frac{\alpha \lambda_i s}{1 + \alpha \lambda_i}\right), \tag{A2.29b}$$

where α is the average received energy-to-noise ratio and the λ_i are the eigenvalues of the complex correlation function. Note that

$$\gamma_1(s) = \gamma_0(s - 1) - \gamma_0(-1). \tag{A2.30}$$

1. Lower Bound

a. Bounds for Distribution Functions: The results of (A3.5) through (A3.7) enable us to express the probabilities $P(x \le h)$ and $P(y \ge h)$ as

$$P(x \le h) = B_0 \exp - [s\gamma_0'(s) - \gamma_0(s)], \tag{A2.31a}$$

$$P(y \ge h) = B_1 \exp - [t\gamma_1'(t) - \gamma_1(t)] \tag{A2.31b}$$

provided that h is defined by the expression

$$h = \gamma_0'(s) + \delta \sqrt{\gamma_0''(s)} \tag{A2.31c}$$

for some values of s and δ and that in addition

$$\gamma_1'(t) + \delta \sqrt{\gamma_1''(t)} = \gamma_0'(s) + \delta \sqrt{\gamma_0''(s)}. \tag{A2.31d}$$

In these expressions the quantities B_0 and B_1 are defined by the relations

$$B_0 = \int_{-\infty}^{\delta} q(y) \exp - sy \sqrt{\gamma_0''(s)} \, dy \tag{A2.32a}$$

and

$$B_1 = \int_{\delta}^{\infty} \hat{q}(y) \exp - ty \sqrt{\gamma_1''(t)} \, dy, \tag{A2.32b}$$

where $q(y)$ and $\hat{q}(y)$ are, respectively, the probability densities defined by the semi-invariant moment-generating functions

$$\mu(r) = - \sum_i [r\rho_i + \ln (1 - r\rho_i)] \tag{A2.33a}$$

and

$$\hat{\mu}(r) = - \sum_i [r\hat{\rho}_i + \ln (1 - r\hat{\rho}_i)] \tag{A2.33b}$$

with

$$\rho_i = [\gamma_0''(s)]^{-1/2} \frac{\alpha \lambda_i}{1 - \alpha \lambda_i s} \tag{A2.34a}$$

and

$$\hat{\rho}_i = [\gamma_1''(t)]^{-1/2} \frac{\alpha \lambda_i}{1 + \alpha \lambda_i (1 - t)}. \tag{A2.34b}$$

By virtue of (A2.30), (A2.31d) is valid if t equals $1 + s$. It also follows from (A2.30), (A2.33), and (A2.34) that the two tilted densities $q(y)$ and $\hat{q}(y)$ are identical if t is set equal to $1 + s$.

Upon setting t equal to $1 + s$ and defining the function

$$\gamma(s) = \frac{1}{\alpha} [\gamma_o(s) + s\gamma_o(-1)], \tag{A2.35}$$

we obtain

$$P(x \le h) = B_0 \exp -\alpha[s\gamma'(s) - \gamma(s)] \tag{A2.36a}$$

and

$$P(y \ge h) = B_1 \exp -\alpha[(1 + s)\gamma'(s) - \gamma(s)] \tag{A2.36b}$$

provided that h is determined by the expression

$$h = \alpha\gamma'(s) - \gamma_o(-1) + \delta\sqrt{\alpha\gamma''(s)}. \tag{A2.36c}$$

The values of the coefficients B_0 and B_1 are now determined by the relations

$$B_0 = \int_{-\infty}^{\delta} q(y) \exp - sy\sqrt{\alpha\gamma''(s)}\, dy \tag{A2.37a}$$

and

$$B_1 = \int_{\delta}^{\infty} q(y) \exp - (1 + s)y\sqrt{\alpha\gamma''(s)}\, dy, \tag{A2.37b}$$

where $q(y)$ is, as before, the probability density associated with the semi-invariant moment-generating function $\mu(r)$ of (A2.33a). For future reference, we restate $\mu(r)$ as

$$\mu(r) = - \sum_i [r\rho_i + \ln (1 - r\rho_i)], \tag{A2.38a}$$

where

$$\rho_i = \frac{1}{\sqrt{\alpha\gamma''(s)}} \frac{\alpha\lambda_i}{1 - \alpha\lambda_i s}, \quad i = 1, \dots. \tag{A2.38b}$$

b. The Exponent. Upon introducing the right members of (A2.36a) and (A2.36b) into (A2.28a), we obtain the inequality

$$P(\varepsilon) \ge \frac{B_0 B_1}{4} m \exp - \alpha[(1 + 2s)\gamma'(s) - 2\gamma(s)] \tag{A2.39a}$$

provided that

$$B_1 m \exp -\alpha[(1 + s)\gamma'(s) - \gamma(s)] \le 1. \tag{A2.39b}$$

We now seek to diminish further the right member of (A2.39a) as so to obtain a simpler, albeit a weaker, result. To this end we note that the validity of the equation is not altered if a lower bound to B_0 and B_1 are substituted for B_0 and B_1. Moreover (A2.39b) will certainly be satisfied if it is satisfied when B_1 is replaced by a larger quantity. The freedom to alter the values of both B_0 and B_1 without invalidating our results is important to us because these quantities often play a secondary role in determining the value of the system error probability—a role that does not warrant determination of their exact value. This is particularly true if the value of s lies in the interval -1 to 0.

The derivation of suitable bounds to the quantities B_0 and B_1 is the central problem considered in Appendix 3. We present the relevant results in Section C1c; for the moment it suffices to state that, by virtue of Section B1a of Appendix 3, B_1 does not exceed unity for values of s in the range $-1 \le s \le 0$. Therefore (A2.39b) is satisfied if

$$m \le \exp \alpha[(1 + s)\gamma'(s) - \gamma(s)] \qquad \text{(A2.40}a)$$

provided that

$$-1 \le s \le 0. \qquad \text{(A2.40}b)$$

Since we are concerned only with values of s that satisfy (A2.40b), we have

$$P(\varepsilon) \ge K_1 m \exp -\alpha[(1 + 2s)\gamma'(s) - 2\gamma(s)] \qquad \text{(A2.41}a)$$

provided that (A2.40a) is satisfied, where

$$K_1 = \tfrac{1}{4}B_0 B_1 \qquad \text{(A2.41}b)$$

with B_0 and B_1 determined by (A2.37a) and (A2.37b).

Our next task is to maximize the bound of (A2.41) with respect to the parameter s. Since the coefficient K_1 is frequently less important than the exponent $-(1 + 2s)\gamma'(s) - 2\gamma(s)$, we maximize the exponent subject to the constraints of (A2.40).

The maximization is greatly simplified by the fact that $\gamma(s)$ is convex in s. In particular, the derivative of the exponent is $-(1 + 2s)\gamma''(s)$; hence the exponent is an increasing function of s for $-1 \le s \le -\tfrac{1}{2}$ and a decreasing function of s for $-\tfrac{1}{2} \le s \le 0$. Since the right member of (A2.40a) is an increasing function of s, s should be set equal to $-\tfrac{1}{2}$ if

$$m \le \exp \alpha[\tfrac{1}{2}\gamma'(-\tfrac{1}{2}) - \gamma(-\tfrac{1}{2})]. \qquad \text{(A2.42)}$$

If m exceeds the right member of (A2.42), s should be chosen to satisfy (A2.40a) with the equality sign provided that such a choice satisfies the constraint of (A2.40b). Since the right member of (A2.40a) is an increasing function of s for $s \ge -1$, such a choice is possible if and only if m exceeds the right member of (A2.42) but does not exceed $\exp \alpha\gamma'(0)$.

The results of the optimization are conveniently presented in the same terms as were the bounds for the additive white Gaussian noise channel. Thus we introduce the information rate R of (A2.17a), the time τ allotted to the transmission of a waveform, and the capacity C of that additive white Gaussian noise channel which has the same average-power–to–noise ratio as the given channel has (A2.17b). The bound of (A2.41) then becomes, after optimization,

$$P(\varepsilon) \geq K_1 \cdot 2^{-\tau CE}, \qquad (A2.43)$$

where the exponent E is related to the information rate and the system design parameters through the parametric equations

$$E = s\gamma'(s) - \gamma(s)$$

with $\qquad\qquad\qquad\qquad\qquad\qquad\qquad\qquad\qquad\qquad$ (A2.44a)

$$\frac{R}{C} = (1 + s)\gamma'(s) - \gamma(s)$$

for $R\mathrm{crit} \leq R \leq \gamma'(0)C$, or equivalently for $-\frac{1}{2} \leq s \leq 0$, and

$$E = -2\gamma(-\tfrac{1}{2}) - \frac{R}{C} \qquad (A2.44b)$$

for $R \leq R\mathrm{crit}$ where $R\mathrm{crit}$, the critical rate, is defined by the expression

$$\frac{R\mathrm{crit}}{C} = \tfrac{1}{2}\gamma'(-\tfrac{1}{2}) - \gamma(-\tfrac{1}{2}). \qquad (A2.44c)$$

The function $\gamma(s)$ is defined by (A2.35), which becomes, when combined with (A2.29),

$$\gamma(s) = -\frac{1}{\alpha}\sum_i [\ln (1 - \alpha\lambda_i s) + s \ln (1 + \alpha\lambda_i)]. \qquad (A2.45)$$

Equations (A2.43) through (A2.45) constitute the basic lower bound to the error probability of a fading dispersive channel. It remains to provide lower bounds for K_1, the product $\frac{1}{4}B_0 B_1$. Such bounds are now determined.

c. The Coefficient. The quantities B_0 and B_1, as introduced in (A2.32a) and (A2.32b), are just the coefficients discussed in Appendix 3. The subsequent modification of B_1, through the substitution $t = 1 + s$, did not alter this equivalence. Thus the values of B_0 and B_1, as given by (A2.37) may be under bounded by an application of the results in Appendix 3. Three sets of estimates are now presented: a general but weak bound, an asymptotic result, and a sharp bound for equal eigenvalues.

The bound of (A3.39) may be applied to B_0 to obtain the following inequality

$$B_0 \geq \tfrac{1}{2}(1 - A^{-2}) \exp - |s| A \sqrt{\alpha \gamma''(s)}, \qquad \text{for } s \leq 0, \qquad \text{(A2.45a)}$$

provided that

$$\int_{-A}^{\delta} q(y) \, dy = \int_{\delta}^{A} q(y) \, dy, \qquad \text{(A2.45b)}$$

where $q(y)$ is the probability density specified by the semi-invariant moment-generating function of (A2.38). In a similar manner, it follows from (A2.37b) and (A3.40) that

$$B_1 \geq \tfrac{1}{2}(1 - A^{-2}) \exp -(1 + s)A \sqrt{\alpha \gamma''(s)}, \qquad \text{for } 1 + s \geq 0, \quad \text{(A2.46)}$$

provided that (A2.45b) is satisfied. Since the value of δ has not yet been specified, it may be chosen, for any value of A, to satisfy (A2.45b). Such a choice is always possible because $q(y)$ is a continuous function of y.

Upon combining (A2.45) and (A2.46) and setting A equal to $\sqrt{2}$, the following lower bound to the product $\tfrac{1}{4} B_0 B_1$ is obtained

$$K_1 = \tfrac{1}{4} B_0 B_1 \geq \tfrac{1}{64} \exp -\sqrt{2\alpha \gamma''(s)}. \qquad \text{(A2.47)}$$

The right member of this expression may be further simplified upon noting that, by virtue of (A2.45),

$$\gamma''(s) = \frac{1}{\alpha} \sum_i \left(\frac{\alpha \lambda_i}{1 - \alpha s \lambda_i} \right)^2,$$

or for negative values of s,

$$\gamma''(s) \leq \alpha \sum_i \frac{(\lambda_i)^2}{2|s| \alpha \lambda_i} = \frac{1}{2|s|}. \qquad \text{(A2.48)}$$

Hence

$$K_1 = \frac{1}{4} B_0 B_1 \geq \frac{1}{64} \exp - \sqrt{\frac{\alpha}{|s|}}. \qquad \text{(A2.49)}$$

We next apply the asymptotic results of Appendix 3, Section B3, to the coefficients B_0 and B_1 as specified by (A2.32a) and (A2.32b). The relevant expressions are provided by (A3.55) and (A3.56). Upon changing the notation to conform to the problem at hand, we obtain the following results, which are valid for $-\tfrac{1}{2} \leq s < 0$:

$$B_0 \sim \frac{1}{|s| \sqrt{2\pi \alpha \gamma''(s)}}, \qquad \text{(A2.50a)}$$

$$B_1 \sim \frac{1}{|1 + s| \sqrt{2\pi \alpha \gamma''(s)}}. \qquad \text{(A2.50b)}$$

The symbol $\sim$ has been employed to denote that ratios of the right and left members of the expressions tend to unity as $\alpha\gamma''(s)$ tends to infinity for any fixed negative value of s in the interval $-\frac{1}{2}$ to 0.

The only noteworthy point about the transition from (A3.55) and (A3.56) to (A2.50a) and (A2.50b) is the following. In the present application the quantities σ_i associated with (A3.55) are given by the expression

$$\sigma_i = \frac{\alpha\lambda_i}{1 + \alpha\lambda_i}.$$

Thus, the σ_i are all less than unity and the constraints associated with (A3.55) are automatically satisfied for all fixed negative values of s between 0 and $-\frac{1}{2}$.

Equations (A2.50a) and (A2.50b) yield the following asymptotic expression for K_1, the product $\frac{1}{4}B_0 B_1$:

$$K_1 \sim \frac{1}{4|s| \, |1 + s|2\pi\alpha\gamma''(s)}, \tag{A2.51a}$$

or by virtue of (A2.48) and the fact that $|1 + s| \leq 1$ for $-\frac{1}{2} \leq s \leq 0$,

$$\lim_{\alpha \to \infty} K_1 \geq \frac{1}{4\pi\alpha}. \tag{A2.51b}$$

We now suppose that only D of the eigenvalues λ_i are positive and that all of the positive eigenvalues are equal to D^{-1}. This supposition, in conjunction with (A2.32a) and (A2.32b) and with (A3.53) and (A3.54), permits us to conclude that

$$B_0 \geq \frac{0.96}{\sqrt{2\pi(1 + D)}}, \tag{A2.52a}$$

$$B_1 \geq \frac{0.96}{\sqrt{2\pi(1 + D)}} \frac{1 + (\alpha/D)|s|}{1 + \alpha/D} \tag{A2.52b}$$

for $-\frac{1}{2} \leq s \leq 0$. Thus K_1 satisfies the inequality

$$K_1 \geq \frac{0.2}{2\pi} \frac{1 + (\alpha/D)|s|}{(1 + D)(1 + \alpha/D)}. \tag{A2.53}$$

In summary, the value of K_1 in the lower bound to the error probability satisfies the inequalities

$$K_1 \geq \tfrac{1}{64} \exp - \sqrt{2\alpha\gamma''(s)} \tag{A2.54a}$$

and

$$K_1 \geq \tfrac{1}{64} \exp - \frac{\sqrt{\alpha}}{|s|}. \tag{A2.54b}$$

Moreover as $\alpha\gamma''(s)$ tends to infinity for a fixed negative value of s, the asymptotic behavior of K_1 is determined from expression

$$K_1 \sim \frac{1}{4|s|(1+s) \cdot 2\pi\alpha\gamma''(s)}. \qquad (A2.54c)$$

Finally, if precisely D of the eigenvalues are nonzero and if these D eigenvalues equal each other,

$$K_1 \geq \frac{0.2}{2\pi} \frac{1+(\alpha/D)|s|}{(1+D)(1+\alpha/D)}. \qquad (A2.54d)$$

Equations A2.43, A2.44, and A2.54 provide the basic lower bounds to the error probability of fading dispersive channels. It remains to show that these bounds are relatively sharp.

2. Upper Bounds

According to (A2.27), (A2.29), and (A2.30), the error probability satisfies the inequality

$$P(\varepsilon) \leq P(x \leq h) + mP(h < x \leq y), \qquad (A2.55)$$

where x and y are statistically independent random variables with the semi-invariant moment-generating functions

$$\gamma_0(s) = -\sum_i \ln (1 - \alpha\lambda_i s) \qquad (A2.56a)$$

and

$$\gamma_1(s) = \gamma_0(s-1) - \gamma_0(-1), \qquad (A2.56b)$$

respectively.

a. Bounds to Distribution Functions: Our first objective is to over bound the probability $P(h < x \leq y)$. This may be done through an application of (A3.90) to obtain

$$P(h < x \leq y) \leq B_2 \exp -[(t+r)\gamma_0'(r) - \gamma_0(r) - \gamma_1(t)] \qquad (A2.57)$$

provided that t is nonnegative and that $h = \gamma_0'(r)$.
The coefficient B_2 is precisely determined by (A3.91a). Equation A3.92 shows that

$$B_2 \leq 1 \qquad (A2.58)$$

provided that $r + t$ is nonnegative. A stronger bound is, by virtue of (A3.93),

$$B_2 \leq \frac{1}{|t+r|\sqrt{2\pi\gamma_0''(r)}} \left(\frac{\eta'}{\eta'-1}\right)^2 \qquad (A2.59)$$

provided that $r + t$ is nonnegative and that r does not exceed $(\alpha\lambda_o)^{-1}$, where λ_o is the maximum value of the λ_i. The quantity η' is determined by the expression

$$\eta' = \gamma_o''(r)\left(r - \frac{1}{\alpha\lambda_o}\right)^2. \tag{A2.60}$$

An upper bound to the probability $P(x \le h)$ is also required. An appropriate bound is provided by (A3.15) and (A3.16). Specifically if s is not positive and if $\gamma_o'(s) = h$, then

$$P(x \le h) = B_0 \exp - [s\gamma_o'(s) - \gamma_o(s)], \tag{A2.61}$$

where B_0 satisfies the inequalities

$$B_0 \le 1, \tag{A2.62a}$$

$$B_0 \le \frac{1}{|s|\sqrt{2\pi\gamma_o''(s)}}\left(\frac{\eta}{\eta - 1}\right)^2 \tag{A2.62b}$$

with

$$\eta = \gamma_o''(s)\left(s - \frac{1}{\alpha\lambda_o}\right)^2. \tag{A2.62c}$$

Equations A2.57 and A2.61 require that r and s be chosen so that $\gamma_o'(s)$ equals $\gamma_o'(r)$. Since $\gamma_o(s)$ is a convex function of s, this implies that r equals s.

b. High Rate Bound: Upon setting r equal to s in (A2.57) and (A2.59) through (A2.62) and combining the result with (A2.55) and (A2.56), we obtain

$$P(\varepsilon) \le k_o\{\exp - [s\gamma_o'(s) - \gamma_o(s)]\}$$
$$\times \{1 + k_1 m \exp - [t\gamma_o'(s) - \gamma_o(t - 1) + \gamma_o(-1)]\} \tag{A2.63}$$

provided that $s + t$ is nonnegative and s does not exceed zero. In this expression k_o and k_1 may either both be set equal to unity or be computed from the expressions

$$k_o \le \frac{1}{|s|\sqrt{2\pi\gamma_o''(s)}}\left(\frac{\eta}{\eta - 1}\right)^2, \tag{A2.64}$$

$$k_1 \le \left|\frac{s}{s + t}\right|. \tag{A2.65}$$

Our next task is to choose the value of the parameter t so as to approximately minimize the right member of (A2.63). Were it not for the dependence of k_1 upon t, the minimum would be achieved by setting t equal to $1 + s$

unless s were less than $-\frac{1}{2}$. For the moment we restrict the value of s to exceed $-\frac{1}{2}$, and since the dependence of k_1 upon t is less important than that of the exponent, we set t equal to $1 + s$. The bound of (A2.63) then may be reduced rather quickly to the same form as the lower bound of (A2.43) and (A2.44). Since the procedure is analogous to that employed with the additive white Gaussian noise channel, it is not repeated here. The result is as follows:

$$P(\varepsilon) \le K_2' \cdot 2^{-tCE}, \qquad (A2.66)$$

where E is defined parametrically by the expression

$$E = s\gamma'(s) - \gamma(s), \qquad (A2.67a)$$

$$\frac{R}{C} = (1 + s)\gamma'(s) - \gamma(s) \qquad (A2.67b)$$

for $R\text{crit} \le R \le \gamma'(0)C$ or equivalently for $-\frac{1}{2} \le s \le 0$, where

$$\frac{R\text{crit}}{C} = \frac{1}{2}\gamma'(-\frac{1}{2}) - \gamma(-\frac{1}{2}). \qquad (A2.68)$$

It follows from (A2.63) through (A2.65) that the coefficient K_2' does not exceed 2 and that it satisfies the inequality

$$K_2' \le \frac{1}{|s|\sqrt{2\pi\alpha\gamma''(s)}} \frac{1 - |s|}{1 - 2|s|} \left(\frac{\eta}{\eta - 1}\right)^2, \qquad (A2.69a)$$

where

$$\eta = \alpha\gamma''(s)\left(s - \frac{1}{\alpha\lambda_o}\right)^2. \qquad (A2.69b)$$

c. Low Rate Bound: Equations A2.66 through A2.69 comprise the basic upper bound to the error probability of a fading dispersive system for rates greater than the critical rate $R\text{crit}$. It remains only to provide an upper bound to $P(\varepsilon)$ for rates less than the critical rate. Although such a bound can be deduced from (A2.63), a stronger and simpler result is obtained by first setting h equal to $-\infty$ in (A2.55); that is,

$$P(\varepsilon) \le mp(x \le y),$$

or

$$P(\varepsilon) \le mP(z \le 0), \qquad (A2.70)$$

where the random variable $z = x - y$ has been introduced. The semi-invariant moment-generating function $\gamma_2(s)$ of this variable is given by the expression

$$\gamma_2(s) = \gamma_o(s) + \gamma_1(-s), \qquad (A2.71a)$$

or by virtue of (A2.56a) and (A2.56b),

$$\gamma_2(s) = -\sum_i \ln\left[(1 - \alpha\lambda_i s)\left(1 + \frac{\alpha\lambda_i s}{1 + \alpha\lambda_i}\right)\right]. \qquad (A2.71b)$$

Upon applying the results of (A3.5a), (A3.15), (A3.16a), and (A3.18b) to (A2.70) and (A2.71b), we obtain

$$P(z \leq 0) = B_3 \exp - [t\gamma_2'(t) - \gamma_2(t)] \qquad (A2.72a)$$

provided that

$$\gamma_2'(t) = 0, \qquad (A2.72b)$$

where, if t does not exceed zero,

$$B_3 \leq 1 \qquad (A2.73a)$$

and

$$B_3 \leq \frac{1}{|t|\sqrt{2\pi\gamma_2''(t)}} \left(\frac{\tilde{\eta}}{\tilde{\eta} - 1}\right)^2. \qquad (A2.73b)$$

The quantity $\tilde{\eta}$ is given by the expression

$$\tilde{\eta} = \min_i \left[\gamma_2''(t)\left(t - \frac{1}{\sigma_i}\right)^2\right], \qquad (A2.74)$$

where the σ_i take on the values $\alpha\lambda_i$ and $-\alpha\lambda_i/(1 + \alpha\lambda_i)$, $i = 1, \ldots$.

It follows from (A2.71) that (A2.72b) is satisfied when t equals $-\frac{1}{2}$. Upon introducing this value of t into (A2.72) through (A2.74) and expressing $\gamma_2(-\frac{1}{2})$ in terms of the function $\gamma(s)$, the bound of (A2.70) becomes

$$P(\varepsilon) \leq K_2'' \cdot 2^{-\tau C \hat{E}} \qquad (A2.75a)$$

with

$$\hat{E} = -2\gamma(-\tfrac{1}{2}) - \frac{R}{C}, \qquad (A2.75b)$$

where

$$K_2'' \leq 1 \qquad (A2.76a)$$

and, also

$$K_2'' \leq \frac{1}{\sqrt{\pi\alpha\gamma''(-\frac{1}{2})}} \left(\frac{\eta_c}{\eta_c - 1}\right)^2 \qquad (A2.76b)$$

with

$$\eta_c = 2\alpha\gamma''\left(-\frac{1}{2}\right)\left(\frac{1}{2} + \frac{1}{\alpha\lambda_o}\right)^2. \qquad (A2.77)$$

The best upper bound to the error probability is obtained by employing (A2.66) for rates greater than the critical rate and (A2.75) for rates less than the critical rate. For ease of reference the results are summarized below.

$$P(\varepsilon) \leq K_2 \cdot 2^{-\tau C E}, \qquad (A2.78)$$

where E is determined by (A2.44) and K_2 may be determined by either of the following two expressions:

$$K_2 \leq \begin{cases} 1, & \text{for} \quad R \leq R\text{crit}, \\ 2, & \text{for} \quad R > R\text{crit}, \end{cases} \tag{A2.79a}$$

$$K_2 \leq \begin{cases} \dfrac{1}{\sqrt{\pi\alpha\gamma''(-\frac{1}{2})}} \left(\dfrac{\eta_c}{\eta_c - 1}\right)^2, & R \leq R\text{crit}, \\[4mm] \dfrac{1}{|s|\sqrt{2\pi\alpha\gamma''(s)}} \dfrac{1 - |s|}{1 - 2|s|} \left(\dfrac{\eta}{\eta - 1}\right)^2, & R\text{crit} \leq R \leq C\gamma'(0), \end{cases} \tag{A2.79b}$$

where $\gamma(s)$ is determined by (A2.45) and

$$\eta_c = 2\alpha\gamma''\left(-\frac{1}{2}\right)\left(\frac{1}{2} + \frac{1}{\alpha\lambda_o}\right)^2, \tag{A2.80a}$$

$$\eta = \alpha\gamma''(s)\left(|s| + \frac{1}{\alpha\lambda_o}\right)^2. \tag{A2.80b}$$

Finally we note that in the special instance of equal eigenvalues

$$\gamma''(s) = \frac{\alpha}{D}\left(1 - \frac{\alpha s}{D}\right)^{-2}, \tag{A2.81}$$

and so (A2.79) becomes

$$K_2 \leq \begin{cases} \dfrac{1}{\sqrt{\pi\alpha}}\left(\dfrac{D}{\alpha}\right)^{1/2}\left(1 + \dfrac{\alpha}{2D}\right)\left(\dfrac{2D}{2D - 1}\right)^2, & R \leq R\text{crit}, \\[4mm] \dfrac{1}{|s|\sqrt{2\pi\alpha}}\left(\dfrac{D}{\alpha}\right)^{1/2}\dfrac{1 - |s|}{1 - 2|s|}\left(1 + \dfrac{\alpha|s|}{D}\right)\left(\dfrac{D}{D - 1}\right)^2, & R > R\text{crit}. \end{cases} \tag{A2.82}$$

REFERENCES

[1] R. M. Fano, *Transmission of Information.* New York: Wiley, and M.I.T. Press, 1961, p. 204.

[2] W. Feller, *An Introduction to Probability Theory and Its Applications,* Vol. I. New York: Wiley, 1968, pp. 174–175.

[3] J. M. Wozencraft and I. M. Jacobs, *Principles of Communication Engineering.* New York: Wiley, 1965, pp. 342–346.

[4] L. H. Zetterburg, "Data Transmission over a Noisy Gaussian Channel." *Trans. Roy. Inst. Technol.,* Stockholm, No. 184, 1961.

Appendix 3

Bounds to the Tails of Distribution Functions

The derivations of Appendix 2 require certain bounds to the tails of probability distribution functions. Those bounds are derived in this appendix. They are obtained by first expressing distribution functions in terms of a "tilted probability density," then separating the desired expression into an "exponent" and a "coefficient," and finally deriving suitable bounds to the coefficients. Part A of the appendix draws heavily upon the presentations of Shannon [1] and Fano [2].

A. THE TILTED PROBABILITY FORMULATION

Let x denote a random variable with the probability density $p(x)$ and the semi-invariant moment-generating function $\gamma(s)$, that is,

$$\gamma(s) = \ln g(s), \qquad (A3.1a)$$

where

$$g(s) = \int p(x) \exp sx \, dx. \qquad (A3.1b)$$

We suppose throughout that $p(x)$ is not "impulsive." This supposition is satisfied for the densities that interest us in Appendix 2 and in the subsequent sections of this appendix. Our ultimate objective is to obtain upper and lower bounds to the probability distribution of x. As a first step in that direction, the distribution is expressed in terms of a tilted probability density, $f(x)$.

The probability density $f(x)$ is defined by the expression

$$f(x) = \frac{p(x) \exp tx}{g(t)}, \qquad (A3.2)$$

234

where the value of t is a parameter. The introduction of this new density into the expression for the distribution function of x yields

$$P(x \leq h) = g(t) \int_{-\infty}^{h} f(x) \exp - tx \, dx. \tag{A3.3a}$$

Similarly

$$P(x \geq h) = g(t) \int_{h}^{\infty} f(x) \exp - tx \, dx. \tag{A3.3b}$$

We next define a normalized random variable y, or equivalently a new variable of integration, by the expression

$$y = \frac{x - \gamma'(t)}{\sqrt{\gamma''(t)}}, \tag{A3.4}$$

where $\gamma'(t)$ and $\gamma''(t)$ are, respectively, the first and second derivatives of $\gamma(t)$ with respect to t. Equation A3.3 then becomes

$$P(x \leq h) = B_0 \exp - [t\gamma'(t) - \gamma(t)] \tag{A3.5a}$$

and

$$P(x \geq h) = B_1 \exp - [t\gamma'(t) - \gamma(t)], \tag{A3.5b}$$

where

$$B_0 = \int_{-\infty}^{\delta} q(y) \exp - ty\sqrt{\gamma''(t)} \, dy, \tag{A3.6a}$$

$$B_1 = \int_{\delta}^{\infty} q(y) \exp - ty\sqrt{\gamma''(t)} \, dy \tag{A3.6b}$$

and where

$$\delta = \frac{h - \gamma'(t)}{\sqrt{\gamma''(t)}}, \tag{A3.6c}$$

$$q(y) = \sqrt{\gamma''(t)} f(\gamma'(t) + y\sqrt{\gamma''(t)}). \tag{A3.6d}$$

It is clear that the quantity $q(y)$ of (A3.6d) can be interpreted as a probability density. In fact it is just the density obtained by normalizing the variable x, described by $f(x)$, to zero mean and unit variance as done in (A3.4). To verify this assertion, we note that the semi-invariant moment-generating function $\mu(s)$ associated with $q(y)$ is given by the expression

$$\mu(s) = \gamma\left(t + \frac{s}{\sqrt{\gamma''(t)}}\right) - s\frac{\gamma'(t)}{\sqrt{\gamma''(t)}} - \gamma(t). \tag{A3.7}$$

Since

$$\bar{y} = \int y q(y)\, dy = \frac{d\mu(s)}{ds}\bigg|_{s=0} = \frac{\gamma'(t)}{\sqrt{\gamma''(t)}} - \frac{\gamma'(t)}{\sqrt{\gamma''(t)}} = 0$$

and since

$$\text{var } y = \int (y - \bar{y})^2 q(y)\, dy = \frac{d^2\mu(s)}{d^2 s}\bigg|_{s=0} = \frac{\gamma''(t)}{\gamma''(t)} = 1,$$

the mean and variance of the random variable described by $q(y)$ are indeed zero and unity, respectively.

B. BOUNDS TO THE COEFFICIENTS

The coefficients B_0 and B_1 are next suitably bounded for the particular conditions of interest in Chapter 5. Specifically we consider variables whose semi-invariant moment-generating functions are of the form

$$\gamma(t) = -\sum_i \ln\,(1 - t\sigma_i). \tag{A3.8}$$

For the most part we suppose that the σ_i are nonnegative.

The semi-invariant moment-generating function $\mu(s)$ associated with the normalized tilted probability density $q(x)$ can be determined from (A3.7) and (A3.8). The result is

$$\mu(s) = -\sum_i [s\rho_i + \ln\,(1 - s\rho_i)], \tag{A3.9}$$

where

$$\rho_i = [\gamma''(t)]^{-1/2}\,\frac{\sigma_i}{1 - \sigma_i t}, \qquad i = 1,\dots. \tag{A3.10}$$

Several properties of the ρ_i should now be noted for future reference. First, the ρ_i are all of the same sign if

$$\sigma_i t < 1, \qquad i = 1,\dots. \tag{A3.11}$$

In particular the ρ_i are all of the same sign if t is less than σ_o^{-1}, where σ_o is the maximum value of the σ_i, i.e.,

$$\sigma_o = \max_i\,(\sigma_i). \tag{A3.12}$$

We consider only values of t for which (A3.11) is satisfied.

Second, the sum of the ρ_i^2 equals unity; that is,

$$\sum_i \rho_i^2 = 1. \tag{A3.13}$$

This may be verified either through the use of the expression for $\gamma''(t)$ or by noting that the left-hand member of (A3.13) is just the variance of the tilted probability density.

Finally, it is easy to verify that

$$\sum_i \rho_i = \frac{\gamma'(t)}{\sqrt{\gamma''(t)}}. \tag{A3.14}$$

The aforestated properties are used in the derivation of upper and then lower bounds to the coefficients B_0 and B_1. We now establish bounds that are applicable to random variables whose semi-invariant moment-generating functions are of the form of (A3.8). We also establish the asymptotic behavior of B_0 and B_1 as $\gamma''(t)$ tends to infinity for fixed values of σ_0 and t.

1. Upper Bounds

We first obtain simple, albeit somewhat weak, upper bounds to the coefficients B_0 and B_1 of (A3.6a) and (A3.6b). Sharper results are obtained subsequently by imposing additional constraints upon the σ_i.

a. Weak Bounds. The value of B_0 does not exceed unity if neither t nor δ is positive. Similarly the value of B_1 does not exceed unity if neither t nor δ is negative. The first assertion is established by noting that

$$B_0 \leq \left[\exp - t\delta\sqrt{\gamma''(t)}\right] \int_{-\infty}^{\delta} q(y)\, dy \leq \int_{-\infty}^{\infty} dy\, q(y) = 1 \tag{A3.15}$$

for nonpositive values of both t and δ. A similar derivation applies to B_1 for nonnegative values of both t and δ. Note that these results are valid no matter what the values of the σ_i.

b. Sharp Bounds. The bound of unity to the coefficients is an exceedingly simple one. It can however be significantly improved at the expense of some complication. In particular if δ is set equal to zero and if the σ_i are nonnegative,

$$B_0 \leq \frac{1}{|t|\sqrt{2\pi\gamma''(t)}} \left(\frac{\eta}{\eta - 1}\right)^2 \tag{A3.16a}$$

provided that

$$t \leq 0 \tag{A3.16b}$$

and

$$B_1 \leq \frac{1}{|t|\sqrt{2\pi\gamma''(t)}} \left(\frac{\eta}{\eta - 1}\right)^2 \tag{A3.17a}$$

provided that

$$0 \leq t < \sigma_0^{-1}, \tag{A3.17b}$$

where σ_o is the maximum value of the σ_i. The quantity η is defined by the expression

$$\eta = [\gamma''(t)]\left(t - \frac{1}{\sigma_o}\right)^2 \tag{A3.18a}$$

More generally (A3.16a) is valid for all nonpositive values of t and (A3.17a) is valid for all nonnegative values of t if η is computed from the expression

$$\eta = \min_i \left[\gamma''(t)\left(t - \frac{1}{\sigma_i}\right)^2\right]. \tag{A3.18b}$$

This is true even if some of the σ_i are negative.

It is shown in Section B3 that these bounds are asymptotically correct in the limit of increasing $\gamma''(t)$ for fixed values of σ_o and t. That is, as $\gamma''(t)$ tends to infinity, the right members of (A3.6a) and (A3.6b) tend to $\left(|t|\sqrt{2\pi\gamma''(t)}\right)^{-1}$ under the stated constraints upon t. The bound of (A3.17) is now derived. The derivation of (A3.16) is essentially identical to that of (A3.17).

By virtue of (A3.6b) with δ set equal to zero,

$$B_1 = \int_0^\infty q(y) \exp - y \Delta \, dy, \tag{A3.19a}$$

where we introduce

$$\Delta = t\sqrt{\gamma''(t)}. \tag{A3.19b}$$

The probability density $q(y)$ is described by the semi-invariant moment-generating function of (A3.9). That equation implies that $q(y)$ is given by the integral expression

$$q(y) = \frac{1}{2\pi}\int_{-\infty}^\infty g(\omega) \exp - j\omega y \, d\omega, \tag{A3.20}$$

where

$$g(\omega) = \prod_i \frac{\exp - j\omega\rho_i}{1 - j\omega\rho_i} \tag{A3.21a}$$

and where, as before,

$$\rho_i = [\gamma''(t)]^{1/2}\frac{\sigma_i}{1 - \sigma_i t} = \frac{1}{\Delta}\frac{\sigma_i t}{1 - \sigma_i t}. \tag{A3.21b}$$

The introduction of (A3.20) into (A3.19a) yields

$$B_1 = \frac{1}{2\pi}\int_0^\infty dy \int_{-\infty}^\infty g(\omega) \exp - y(j\omega + \Delta) \, d\omega. \tag{A3.22}$$

Since the value of Δ is nonnegative, the order of integration may be interchanged and the integral with respect to y may be evaluated first. The result is

$$B_1 = \frac{1}{2\pi} \int_{-\infty}^{\infty} \frac{g(\omega)}{\Delta + j\omega} \, d\omega \qquad (A3.23)$$

provided that $t \geq 0$.

We next observe that, since B_1 is nonnegative,

$$B_1 \leq \frac{1}{2\pi} \int_{-\infty}^{\infty} \frac{|g(\omega)|}{|\Delta + j\omega|} \, d\omega = \frac{1}{2\pi} \int_{-\infty}^{\infty} \frac{|g(\omega)|}{\sqrt{\Delta^2 + \omega^2}} \, d\omega,$$

or since $\Delta^2 + \omega^2 \geq \Delta^2$,

$$B_1 \leq \frac{1}{2\pi \Delta} \int_{-\infty}^{\infty} |g(\omega)| \, d\omega. \qquad (A3.24)$$

Also, by virtue of (A3.21a),

$$|g(\omega)|^2 = \prod_i [1 + (\rho_i \omega)^2]^{-1},$$

or

$$|g(\omega)|^2 = \exp - \sum_i \rho_i^2 \left\{ \frac{\ln [1 + (\rho_i \omega)^2]}{\rho_i^2} \right\}. \qquad (A3.25)$$

It is readily verified that the function in braces in (A3.25) is monotone decreasing in ρ_i^2. Thus

$$\frac{\ln [1 + (\rho_i \omega)^2]}{\rho_i^2} \geq \frac{\ln (1 + \eta^{-1} \omega^2)}{\eta^{-1}} \qquad (A3.26)$$

where η^{-1} is the maximum value of the ρ_i^2. Since, by virtue of (A3.13), the sum of the ρ_i^2 is unity, we have, from (A3.25) and (A3.26),

$$|g(\omega)|^2 \leq \exp - \frac{\ln (1 + \eta^{-1} \omega^2)}{\eta^{-1}},$$

or equivalently

$$|g(\omega)|^2 \leq \left(1 + \frac{\omega^2}{\eta} \right)^{-\eta}. \qquad (A3.27)$$

Upon combining (A3.24) and (A3.27), we obtain

$$B_1 \leq \frac{1}{2\pi \Delta} \int_{-\infty}^{\infty} \left(1 + \frac{\omega^2}{\eta} \right)^{-\eta/2} \, d\omega, \qquad (A3.28)$$

or by a change in the dummy variable of integration,

$$B_1 \leq \frac{\sqrt{\eta}}{2\pi \Delta} \int_{-\infty}^{\infty} (1 + x^2)^{-\eta/2} \, dx.$$

It follows from (A3.13) that the value of η equals or exceeds unity. Consequently the last integral may be evaluated [3] to obtain

$$B_1 \le \frac{1}{2\Delta} \sqrt{\frac{\eta}{\pi}} \frac{\Gamma[(\eta - 1)/2]}{\Gamma(\eta/2)}, \tag{A3.29}$$

where $\Gamma(x)$ is the gamma function. The bound may be further reduced by noting that [3]

$$1 \le \frac{\Gamma(x)}{\sqrt{2\pi} \, x^{(x - 1/2)} \exp - x} \le 1 + \frac{1}{12x}. \tag{A3.30}$$

Upon overestimating $\Gamma[(\eta - 1)/2]$ and underestimating $\Gamma(\eta/2)$, through the use of (A3.30), we obtain

$$B_1 \le \frac{[\exp \tfrac{1}{2}]}{\sqrt{2\pi} \, \Delta} \left[1 + \frac{1}{6(\eta - 1)} \right] (1 - \eta^{-1})^{\eta/2 - 1},$$

or since $(1 - \eta^{-1})^{\eta/2}$ does not exceed $\exp - \tfrac{1}{2}$ and $6(\eta - 1)$ exceeds $\eta - 1$,

$$B_1 \le \frac{1}{\sqrt{2\pi} \, \Delta} \left(\frac{\eta}{\eta - 1} \right)^2. \tag{A3.31}$$

Finally, we recall that, from (A3.21b),

$$\rho_i = \frac{1}{\Delta} \frac{t\sigma_i}{1 - t\sigma_i}.$$

Thus if t does not exceed σ_o^{-1}, where σ_o is the maximum value of all the σ_i, and if the σ_i are nonnegative,

$$\max_i \rho_i = \frac{1}{\Delta} \frac{t\sigma_o}{1 - t\sigma_o}, \qquad i = 1, \dots, \tag{A3.32}$$

and η^{-1}, the maximum value of the ρ_i^2, becomes

$$\eta^{-1} = \left(\frac{1}{\Delta} \right)^2 \left(\frac{t\sigma_o}{1 - t\sigma_o} \right)^2. \tag{A3.33a}$$

More generally if t is nonnegative,

$$\eta^{-1} = \max_i \left[\left(\frac{1}{\Delta} \right)^2 \left(\frac{t\sigma_i}{1 - t\sigma_i} \right)^2 \right]. \tag{A3.33b}$$

Equations A3.17 and A3.18 follow from (A3.31) and (A3.33) upon eliminating Δ by use of (A3.19b).

2. Lower Bounds

We now derive lower bounds to the coefficients B_0 and B_1. The general lower bounds are relatively simple, but they are not as sharp as the upper bounds. Therefore we also consider the special situation in which all of the nonzero σ_i are equal and positive.

a. **Weak Bounds:** The bound to B_0 is obtained by recalling, from (A3.6*a*), that

$$B_0 = \int_{-\infty}^{\delta} q(y) \exp - ty\sqrt{\gamma''(t)}\, dy.$$

Hence

$$B_0 \geq \int_{-A}^{\delta} q(y) \exp - ty\sqrt{\gamma''(t)}\, dy, \qquad (A3.34)$$

where A is any number that exceeds $-\delta$. For nonpositive values of t, the right member of (A3.34) may be further reduced to obtain

$$B_0 \geq [\exp - |t|A\sqrt{\gamma''(t)}] \int_{-A}^{\delta} q(y)\, dy. \qquad (A3.35)$$

We next anticipate the use we make of these bounds. Specifically we choose the value of δ so that

$$\int_{-A}^{\delta} q(y)\, dy = \int_{\delta}^{A} q(y)\, dy = \tfrac{1}{2}Q(|y| \leq A). \qquad (A3.36)$$

Since the random variables we are considering are continuous valued, such a choice is always possible. Subject to this supposition, (A3.35) becomes

$$B_0 \geq \tfrac{1}{2}Q(|y| \leq A) \exp - |t|A\sqrt{\gamma''(t)}. \qquad (A3.37)$$

Finally, we note that $Q(|y| \leq A)$ may be interpreted as the probability that the random variable described by $q(y)$ lies in the interval $-A, A$. Thus $Q(|y| \leq A)$ may be under bounded by an application of Tchebycheff's inequality to obtain

$$Q(|y| \leq A) \geq 1 - A^{-2}, \qquad (A3.38)$$

where we have made use of the fact that $q(y)$ has a mean of zero and a variance of unity.

Upon combining (A3.37) and (A3.38), we obtain

$$B_0 \geq \tfrac{1}{2}(1 - A^{-2}) \exp - |t| A \sqrt{\gamma''(t)} \qquad \text{for } t \leq 0 \qquad \text{(A3.39)}$$

provided that (A3.36) is satisfied. In a similar manner we find that

$$B_1 \geq \tfrac{1}{2}(1 - A^{-2}) \exp - |t| A \sqrt{\gamma''(t)}, \qquad \text{for } t \geq 0, \qquad \text{(A3.40)}$$

provided that (A3.36) is satisfied.

The right members of (A3.39) and (A3.40) are positive only if the value of A exceeds unity. Thus as $\sqrt{\gamma''(t)}$ tends to infinity, the lower bounds to B_0 and B_1 vanish exponentially with $\sqrt{\gamma''(t)}$ in marked contrast to the upper bounds which vanish as $[\gamma''(t)]^{-1/2}$. Although the exponential vanishing with $\sqrt{\gamma''(t)}$ is sometimes acceptable, sharper estimates can be obtained under certain conditions. We turn now to one such set of conditions.

b. Equal σ_i's: We now suppose that D of the σ_i equal σ_0 and that the remaining σ_i are zero. The right member of (A3.9a) then becomes

$$\mu(s) = -D[s\rho_1 + \ln(1 - s\rho_1)], \qquad \text{(A3.41a)}$$

where

$$\rho_1 = [\gamma''(t)]^{-1/2} \frac{\sigma_0}{1 - \sigma_0 t} = \frac{1}{\sqrt{D}}. \qquad \text{(A3.41b)}$$

We suppose that $\sigma_0 t$ is less than unity. This implies that the normalized "tilted" random variable possesses the probability density

$$q(y) = \frac{D^{D/2}}{(D-1)!}(y + \sqrt{D})^{D-1} \exp - \sqrt{D}(y + \sqrt{D}), \quad y \geq -\sqrt{D},$$

$$q(y) = 0, \qquad\qquad\qquad\qquad\qquad\qquad\qquad\qquad \text{elsewhere.}$$

$$\text{(A3.42)}$$

Upon introducing the right member of (A3.42) into the right members of (A3.6a) and (A3.6b), setting δ to zero, and performing some elementary manipulations we obtain the following expressions

$$B_0 = \left[\frac{\exp t \sqrt{D\gamma''(t)}}{(D-1)!}\right] \int_0^D \left(x^{D-1} \exp - x \left\{1 + t\left[\frac{\gamma''(t)}{D}\right]^{1/2}\right\}\right) dx, \quad \text{(A3.43a)}$$

$$B_1 = \left[\frac{\exp t \sqrt{D\gamma''(t)}}{(D-1)!}\right] \int_D^\infty \left(x^{D-1} \exp - x \left\{1 + t\left[\frac{\gamma''(t)}{D}\right]^{1/2}\right\}\right) dx, \quad \text{(A3.43b)}$$

or upon evaluating the integrals,

$$B_0 = \left(\left\{1 + t\left[\frac{\gamma''(t)}{D}\right]^{1/2}\right\}^{-D} \exp - D\right)Q_0 \qquad (A3.44a)$$

and

$$B_1 = \left(\left\{1 + t\left[\frac{\gamma''(t)}{D}\right]^{1/2}\right\}^{-D} \exp - D\right)Q_1, \qquad (A3.44b)$$

where

$$Q_0 = \sum_{i=D}^{\infty} \frac{D^i}{i!}\left\{1 + t\left[\frac{\gamma''(t)}{D}\right]^{1/2}\right\}^i \qquad (A3.45a)$$

and

$$Q_1 = \sum_{i=0}^{D-1} \frac{D^i}{i!}\left\{1 + t\left[\frac{\gamma''(t)}{D}\right]^{1/2}\right\}^i. \qquad (A3.45b)$$

We now seek lower bounds to Q_0 and Q_1. We first consider Q_0 and restrict the value of t to exceed $-\sqrt{D/\gamma''(t)}$ but not 0. Then the summand of (A3.45a) is nonnegative and the value of Q_0 exceeds the value of any element of the sum. In particular

$$Q_0 \geq \frac{D^D}{D!}\left\{1 + t\left[\frac{\gamma''(t)}{D}\right]^{1/2}\right\}^D. \qquad (A3.46)$$

Consequently

$$B_0 \geq \frac{D^D}{D!} \exp - D \qquad (A3.47a)$$

provided that

$$-\left[\frac{D}{\gamma''(t)}\right]^{1/2} \leq t \leq 0. \qquad (A3.47b)$$

The right and left members of (A3.47) may be further simplified by an application of (A3.30). The result is, after some simplification,

$$B_0 \geq \frac{0.96}{\sqrt{2\pi}} \frac{1}{\sqrt{1+D}} \qquad (A3.48)$$

provided, again, that (A3.47b) is satisfied. Finally, we observe that, subject to the supposition of equal-valued σ_i,

$$\gamma''(t) = D\left(\frac{\sigma_0}{1 - \sigma_0 t}\right)^2$$

so (A3.47b) is satisfied for all negative values of t.

Our next task is to lower bound the quantity Q_1 defined by (A3.45b) subject to the restriction that the value of t be positive but less than σ_0^{-1}. A suitable bound is again obtained through the use of a single term in the summation of (A3.45b). In particular

$$Q_1 \geq \frac{D^{D-1}}{(D-1)!}\left\{1 + t\left[\frac{\gamma''(t)}{D}\right]^{1/2}\right\}^{D-1} \tag{A3.49}$$

Consequently

$$B_1 \geq \left(\frac{D^D}{D!}\exp - D\right)\left\{1 + t\left[\frac{\gamma''(t)}{D}\right]^{1/2}\right\}^{-1} \tag{A3.50}$$

Upon over bounding the factorial function by the use of (A3.30), (A3.50) becomes

$$B_1 \geq \frac{0.96}{\sqrt{2\pi}}\frac{1 - \sigma_0 t}{\sqrt{1 + D}} \tag{A3.51}$$

provided that t is nonnegative.

It is instructive to compare the lower bounds of (A3.48) and (A3.51) with the upper bounds that result from (A3.16) and (A3.17) when there are D positive σ_i, each equal to σ_0. Under these conditions

$$\gamma''(t) = D\left(t - \frac{1}{\sigma_0}\right)^{-2} \tag{A3.52a}$$

and

$$\eta = D. \tag{A3.52b}$$

Consequently

$$0.96\left(\frac{D}{D+1}\right)^{1/2} \leq B_0\sqrt{2\pi D} \leq \left(1 - \frac{1}{\sigma_0 t}\right)\left(\frac{D}{D-1}\right)^2 \tag{A3.53}$$

provided that t does not exceed zero, and

$$0.96\left(\frac{D}{D+1}\right)^{1/2} \leq B_1\frac{\sqrt{2\pi D}}{1 - \sigma_0 t} \leq \frac{1}{t\sigma_0}\left(\frac{D}{D-1}\right)^2 \tag{A3.54a}$$

provided that

$$0 \leq \sigma_0 t < 1. \tag{A3.54b}$$

Equations A3.53 and A3.54 clearly indicate that the upper and lower bounds to B_0 and B_1 are quite sharp for large values of $\gamma''(t)$ when only D of the σ_i are positive and each of them equals σ_0. However we still lack sharp bounds when the positive σ_i are not all equal. Therefore we next consider the behavior of B_0 and B_1 when $\gamma''(t)$ is large without constraining the values of the σ_i.

3. Asymptotic Expressions for the Coefficients

Asymptotic expressions for B_0 and B_1 are derived here under the supposition that δ equals zero. For simplicity we suppose that the σ_i are nonnegative. It is shown that B_0 and B_1 approach $\left[|t|\sqrt{2\pi\gamma''(t)}\right]^{-1}$ as $\gamma''(t)$ tends to infinity. More precisely, if σ_0 is the maximum value of σ_i in (A3.8) and if Δ denotes $t\sqrt{\gamma''(t)}$, then

$$\lim_{\Delta \to \infty} [\Delta B_1] = \frac{1}{\sqrt{2\pi}} \qquad (A3.55a)$$

provided that δ equals zero, that

$$0 \le t\sigma_o \le 1, \qquad (A3.55b)$$

and that $t\sigma_o$ does not tend to unity. Also

$$\lim_{\Delta \to \infty} [|\Delta|B_0] = \frac{1}{\sqrt{2\pi}} \qquad (A3.56a)$$

provided that δ equals zero and that

$$t \le 0. \qquad (A3.56b)$$

Since the derivations of (A3.55) and (A3.56) are quite similar, only that of (A3.55) is presented.

By virtue of (A3.19), (A3.21), and (A3.23)

$$\Delta B_1 = \frac{1}{2\pi} \int \frac{\Delta}{\Delta + j(\omega)} g(\omega) \, d\omega, \qquad (A3.57)$$

where

$$\Delta = t\sqrt{\gamma''(t)}, \qquad (A3.58a)$$

$$g(\omega) = \prod_i \frac{\exp - j\omega\rho_i}{1 - j\omega\rho_i}, \qquad (A3.58b)$$

and

$$\rho_i = \frac{1}{\Delta} \frac{t\sigma_i}{1 - t\sigma_i}. \qquad (A3.58c)$$

The integral of (A3.57) is next represented in the form

$$\Delta \int \frac{g(\omega)}{\Delta + j(\omega)} \, d\omega = R_1 + R_2, \tag{A3.59}$$

where

$$R_1 = \Delta \int_{-\sqrt{\eta}/4}^{\sqrt{\eta}/4} \frac{g(\omega)}{\Delta + j\omega} \, d\omega \tag{A3.60a}$$

and

$$R_2 = \Delta \int_{\sqrt{\eta}/4}^{\infty} \frac{g(\omega)}{\Delta + j\omega} \, d\omega + \Delta \int_{-\infty}^{-\sqrt{\eta}/4} \frac{g(\omega)}{\Delta + j\omega} \, d\omega \tag{A3.60b}$$

where η^{-1} is the maximum value of the ρ_i^2. It is now shown that R_2 tends to zero and R_1 tends to $\sqrt{2\pi}$ as Δ tends to infinity.

Clearly, R_2 satisfies the inequality

$$|R_2| \leq |\Delta| \int_{\sqrt{\eta}/4}^{\infty} \left| \frac{g(\omega)}{\Delta + j\omega} \right| d\omega + |\Delta| \int_{-\infty}^{-\sqrt{\eta}/4} \left| \frac{g(\omega)}{\Delta + j\omega} \right| d\omega; \tag{A3.61}$$

or by the symmetry of $|g(\omega)/(\Delta + j\omega)|$ in conjunction with (A3.27),

$$|R_2| \leq 2 \int_{\sqrt{\eta}/4}^{\infty} \left[1 + \left(\frac{\omega}{\Delta} \right)^2 \right]^{-1/2} \left(1 + \frac{\omega^2}{\eta} \right)^{-\eta/2} d\omega. \tag{A3.62}$$

Upon changing the variable of integration, (A3.62) becomes

$$|R_2| \leq \sqrt{\eta} \int_{1/16}^{\infty} \frac{(1 + x)^{-\eta/2}}{\sqrt{x + \eta(x/\Delta)^2}} \, dx, \tag{A3.63}$$

or since the denominator of the integrand exceeds $\sqrt{\eta}/16\Delta$ over the range of integration,

$$|R_2| \leq 16 \, \Delta \int_{1/16}^{\infty} (1 + x)^{-\eta/2} \, dx. \tag{A3.64}$$

This integral can be evaluated, if η exceeds 2, to obtain

$$|R_2| \leq \frac{16 \, \Delta}{(\eta/2 - 1)} \left(1 + \frac{1}{16} \right)^{1-\eta/2} \leq \frac{16 \, \Delta}{\eta/2 - 1}. \tag{A3.65}$$

By virtue of (A.33a),

$$\eta = \Delta^2 \left(1 - \frac{1}{t\sigma_0} \right)^2 \tag{A3.66a}$$

provided that

$$0 \leq t\sigma_0 \leq 1, \tag{A3.66b}$$

where σ_0 is the maximum value of the σ_i.

Clearly the value of η exceeds 2, and (A3.65) is valid, as Δ tends to infinity, if $t\sigma_o$ does not tend to unity. Hence

$$\lim_{\Delta \to \infty} |R_2| \leq \lim_{\Delta \to \infty} \frac{16\,\Delta}{\eta/2 - 1} = 0 \tag{A3.67}$$

provided that $0 \leq t\sigma_o \leq 1$ and that $t\sigma_o$ does not tend to unity with increasing Δ.

It remains to estimate the integral R_1 of (A3.60a). As a first step in that direction we observe that, by virtue of Taylor's theorem [4],

$$\frac{\exp - j\omega\rho_i}{1 - j\omega\rho_i} = \exp - \left[\frac{(\omega\rho_i)^2}{2} - r_i \right] \tag{A3.68}$$

provided that the magnitude of $\omega\rho_i$ is less than unity. The "remainder term" r_i satisfies the inequality

$$|r_i| \leq \frac{|\omega\rho_i|^3}{3} \frac{1}{|1 - |\rho_i\omega||^3}. \tag{A3.69}$$

By virtue of (A3.21) and (A3.68), $g(\omega)$ may be expressed as

$$g(\omega) = \left[\exp - \frac{\omega^2}{2} \right] \exp r, \tag{A3.70}$$

where use has been made of the fact that

$$\sum_i \rho_i^2 = 1 \tag{A3.71a}$$

and where r has been defined as

$$r = \sum_i r_i. \tag{A3.71b}$$

Since the ρ_i are positive and $|\omega|$ does not exceed $1/\rho_o$ the quantity r satisfies the inequalities

$$|r| \leq \sum_i |r_i| \leq \frac{|\omega|^3}{3} \sum_i \rho_i^2 \frac{\rho_o}{|1 - |\rho_o\omega||^3} = \frac{|\omega|^3}{3} \frac{\rho_o}{|1 - |\rho_o\omega||^3}, \tag{A3.72}$$

where ρ_o is the maximum value of the ρ_i.

We next observe that

$$|1 - \exp r| \leq |r| \exp |r|.$$

Consequently (A3.70) implies that

$$\left| g(\omega) - \exp -\frac{\omega^2}{2} \right| \leq |r| \exp \left(|r| - \frac{\omega^2}{2} \right), \tag{A3.73}$$

or by virtue of (A3.72),

$$\left| g(\omega) - \exp -\frac{\omega^2}{2} \right| \leq \frac{1}{3} \frac{|\omega^3 \rho_0|}{|1 - |\rho_0 A||^3} \exp -\frac{\omega^2}{2} \left(1 - \frac{2A\rho_0}{3|1 - |\rho_0 A||^3} \right)$$

provided that

$$|\omega| \leq A \leq 1/\rho_0.$$

Upon taking the value of A to be $(4\rho_0)^{-1}$ and expressing ρ_0 in terms of η, as determined by (A3.33a),

$$\left| g(\omega) - \exp -\frac{\omega^2}{2} \right| \leq \frac{c}{\sqrt{\eta}} |\omega|^3 \exp -\frac{\omega^2}{2} \left(1 - \frac{c}{2} \right) \tag{A3.74a}$$

with

$$c = \frac{4^3}{3^4} \tag{A3.74b}$$

provided that $|\omega|$ does not exceed $\sqrt{\eta}/4$.

Returning now to the expression of (A3.60a), for R_1, we observe that

$$R_1 = \Delta \int_{-\sqrt{\eta}/4}^{\sqrt{\eta}/4} \frac{\exp(-\omega^2/2)}{\Delta + j\omega} d\omega + \Delta \int_{-\sqrt{\eta}/4}^{+\sqrt{\eta}/4} \frac{g(\omega) - \exp(-\omega^2/2)}{\Delta + j\omega} d\omega. \tag{A3.75}$$

The rightmost of these integrals is denoted by R_3. It satisfies the inequality

$$|R_3| \leq \int_{-\sqrt{\eta}/4}^{+\sqrt{\eta}/4} \left| g(\omega) - \exp -\frac{\omega^2}{2} \right| d\omega, \tag{A3.76}$$

or by virtue of (A3.74),

$$|R_3| \leq \frac{2c}{\sqrt{\eta}} \int_0^{\sqrt{\eta}/4} \omega^3 \exp -\frac{\omega^2}{2} \left(1 - \frac{c}{2} \right) d\omega \leq \frac{4c}{(1 - c/2)^2} \frac{1}{\sqrt{\eta}} \int_0^\infty x \exp - x \, dx.$$

Evaluating the integral yields

$$|R_3| \leq \frac{4c}{(1 - c/2)^2} \frac{1}{\sqrt{\eta}}. \tag{A3.77}$$

Hence R_3 tends to zero with increasing η, or by virtue of the discussion involving (A3.67), R_3 tends to zero as Δ tends to infinity provided that $0 \leq \sigma_0 t \leq 1$ and that $\sigma_0 t$ does not tend to unity.

The remaining integral to evaluate in (A3.75) is denoted by R_4. It can be cast in the form

$$R_4 = 2 \int_0^{\sqrt{\eta}/4} \frac{\exp(-\omega^2/2)}{1 + (\omega/\Delta)^2} \, d\omega. \tag{A3.78}$$

Adequate bounds to this expression may be obtained by noting that

$$1 \leq 1 + \left(\frac{\omega}{\Delta}\right)^2 \leq \exp\left(\frac{\omega}{\Delta}\right)^2.$$

Hence

$$2 \int_0^{\sqrt{\eta}/4} \exp -\frac{\omega^2}{2} \left(1 + \frac{2}{\Delta^2}\right) d\omega \leq R_4 \leq 2 \int_0^{\sqrt{\eta}/4} \exp -\frac{\omega^2}{2} \, d\omega. \tag{A3.79}$$

The final upper bound to R_4 is obtained by increasing the upper limit of integration to infinity. The resulting bound is

$$R_4 \leq \sqrt{2\pi}. \tag{A3.80}$$

The lower bound is reduced by expressing the leftmost member of (A3.79) in the form

$$\left(\frac{2\pi}{1 + 2/\Delta^2}\right)^{1/2} [1 - 2\Phi(-\tfrac{1}{4}\sqrt{\eta}\sqrt{1 + 2/\Delta^2})],$$

where $\Phi(x)$ is the probability distribution function of a zero-mean unit-variance Gaussian random variable. Since [5]

$$\Phi(-|x|) \leq \frac{\exp-(x^2/2)}{\sqrt{2\pi}|x|},$$

the quantity R_4 satisfies the inequality

$$R_4 \geq \left(\frac{2\pi}{1 + 2/\Delta^2}\right)^{1/2} \left[1 - \frac{8}{\sqrt{1 + 2/\Delta^2}\sqrt{2\pi\eta}} \exp -\frac{\eta}{16}(1 + 2/\Delta^2)\right]. \tag{A3.81}$$

It follows immediately from (A3.80) and (A3.81) that R_4 tends to $\sqrt{2\pi}$ as Δ, and hence η, tends to infinity. Since R_3 tends to zero under the same limit, R_1 tends to $\sqrt{2\pi}$. Moreover R_2 tends to zero, and so $R_1 + R_2$ tends to $\sqrt{2\pi}$. Therefore by virtue of (A3.57) and (A3.59), the limiting value of ΔB_1 is $1/\sqrt{2\pi}$ as was to be proved.

Since the derivation of the limiting expression for B_0 differs only trivially from that presented above, it is not presented.

C. JOINT DISTRIBUTION FUNCTIONS

It is necessary in Appendix 2 to upper bound a joint probability distribution. Specifically, an upper bound is required to the probability $P(h < x \leq y)$ that x exceeds h but not y, where x and y are statistically independent random variables. A suitable bound to $P(h < x \leq y)$ is now derived.

Let $\gamma_0(s)$ and $\gamma_1(s)$ be, respectively, the semi-invariant moment-generating functions of x and y. Further let their probability densities be $p_0(x)$ and $p_1(y)$. Clearly

$$P(h < x \leq y) = \int_h^\infty p_0(x)\, dx \int_x^\infty p_1(y)\, dy. \qquad (A3.82a)$$

Equivalently

$$P(h < x \leq y) = \int_h^\infty p_0(x)\, dx \int_x^\infty p_1(y)(\exp ty)\exp - ty\, dy$$

$$(A3.82b)$$

for any value of t.

Let us now restrict the value of t to be nonnegative so that the value of $-ty$ does not exceed $-tx$ for values of y greater than x. Consequently

$$P(h < x \leq y) \leq \int_h^\infty p_0(x)\exp - tx\, dx \int_x^\infty p_1(y)\exp ty\, dy.$$

$$(A3.83)$$

The right member of this expression is further increased if the integration over y is extended from $-\infty$ to $+\infty$; that is,

$$P(h < x \leq y) \leq [\exp \gamma_1(t)]\int_h^\infty p_0(x)\exp - tx\, dx, \qquad (A3.84a)$$

where

$$\gamma_1(t) = \ln g_1(t)$$

and

$$g_1(t) = \int_{-\infty}^\infty p_1(y)\exp ty\, dy. \qquad (A3.84b)$$

Next define a new probability density $f(x)$ by the expression

$$f(x) = \frac{p_0(x)\exp - tx}{\exp \gamma_0(-t)}. \qquad (A3.85a)$$

The semi-invariant moment-generating function $\varphi(s)$ of $f(x)$ is

$$\varphi(s) = \gamma_0(s - t) - \gamma_0(-t). \qquad (A3.85b)$$

The introduction of the right member of (A3.85a) in (A3.84a) yields

$$P(h < x \le y) \le \{\exp\left[\gamma_1(t) + \gamma_o(t)\right]\} \int_h^\infty f(x)\, dx. \qquad (A3.86)$$

Upon applying the results of Section A to the integral $\int_h^\infty f(x)\, dx$, we obtain

$$P(h < x \le y) \le \{\exp\left[\gamma_1(t) + \gamma_o(-t)\right]\} B_2 \exp -[\hat{r}\varphi'(\hat{r}) - \varphi(\hat{r})], \qquad (A3.87a)$$

where

$$h = \varphi'(\hat{r}) \qquad (A3.87b)$$

and

$$B_2 = \int_0^\infty q(x) \exp - \hat{r}x\sqrt{\varphi''(\hat{r})}\, dx. \qquad (A3.88)$$

The normalized tilted probability density $q(x)$ is described by the semi-invariant moment-generating function $\mu(v)$ defined as

$$\mu(v) = \varphi\left(\frac{v}{\sqrt{\varphi''(\hat{r})}} + \hat{r}\right) - v\,\frac{\varphi'(\hat{r})}{\sqrt{\varphi''(\hat{r})}} - \varphi(\hat{r}). \qquad (A3.89)$$

Finally upon using (A3.85b) to eliminate $\varphi(\hat{r})$ from (A3.87) through (A3.89), we obtain

$$P(h < x \le y) \le B_2 \exp -[(t + r)\gamma_o'(r) - \gamma_o(r) - \gamma_1(t)] \qquad (A3.90a)$$

for $t \ge 0$, where

$$h = \gamma_o'(r) \qquad (A3.90b)$$

and where we define

$$r = \hat{r} - t.$$

The coefficient B_2 may be expressed as

$$B_2 = \int_0^\infty q(x) \exp - x(t + r)\sqrt{\gamma_o''(r)}\, dx, \qquad (A3.91a)$$

where the semi-invariant moment-generating function of $q(x)$ is

$$\mu(v) = \gamma_o\left(\frac{v}{\sqrt{\gamma_o''(r)}} + r\right) - v\,\frac{\gamma_o'(r)}{\sqrt{\gamma_o''(r)}} - \gamma_o(r). \qquad (A3.91b)$$

It is clear that the expression for B_2 is of the same form as that for B_1 (A3.6b) with δ set to zero. The only relevant difference is that $t\sqrt{\gamma''(t)}$ in the latter expression is replaced by $(t + r)\sqrt{\gamma_o''(r)}$ in the former one. Moreover the semi-invariant moment-generating functions that specify the densities

$q(\cdot)$ of (A3.6) and (A3.91) differ only in the substitution of $\gamma_0(\cdot)$ for $\gamma(\cdot)$ in the bounds for B_2 and of r for t. Thus the upper bounds to B_1 can be transformed into bounds for B_2. For example, the value of B_1 does not exceed unity if δ is taken to be zero and s is nonnegative. Similarly

$$B_2 \le 1 \qquad\qquad (A3.92a)$$

provided that

$$r + t \ge 0. \qquad\qquad (A3.92b)$$

Alternately if $\gamma_0(\cdot)$ is of the form given by (A3.8), the bound of (A3.17a) may be transformed to obtain

$$B_2 \le \frac{1}{|t + r|\sqrt{2\pi\gamma_0''(r)}} \left(\frac{\eta}{\eta - 1}\right)^2 \qquad\qquad (A3.93a)$$

provided that $0 \le r + t$ and that $r < \sigma_0^{-1}$, where σ_0 is the maximum value of the σ_i appearing in (A3.8). The quantity η is defined by (A3.18) as

$$\eta = \gamma_0''(r)\left(r - \frac{1}{\sigma_0}\right)^2. \qquad\qquad (A3.93b)$$

It should be noted that the value of t is here restricted only to being nonnegative. In most applications t is adjusted so as to minimize the value of the bound. Since this optimization depends upon the detailed character of $\gamma_1(\cdot)$ and $\gamma_0(\cdot)$, it is performed in Appendix 2 for the particular problem of interest.

REFERENCES

[1] C. E. Shannon, Unpublished Seminar Notes, Summer 1956, M.I.T.
[2] R. M. Fano, *Transmission of Information*. New York: Wiley, and M.I.T. Press, 1961, Chapter 8.
[3] W. Magnus and F. Oberhettinger, *Formulas and Theorems for the Functions of Mathematical Physics*. New York: Chelsea, 1954, p. 4.
[4] I. S. Sokolnikoff and R. M. Redheffer, *Mathematics of Physics and Modern Engineering*. New York: McGraw-Hill, 1958, p. 144.
[5] W. Feller, *An Introduction to Probability Theory and Its Applications*, Vol. I. New York: Wiley, 1968, pp. 174–175.

Asymptotic Behavior of Eigenvalues

We derive here the asymptotic results of Chapter 6 that are encompassed by (6.45), (6.58), (6.72), (6.76), (6.85), and (6.91) through (6.93).

A. HIGHLY SINGLY DISPERSIVE CHANNELS

In this section we suppose that

$$u(t) = \frac{1}{\sqrt{T}} u_o\left(\frac{t}{T}\right),$$

where

$$\int |u_o(t)|^2 \, dt = \int |u_o(t)|^4 \, dt = 1,$$

and consider the behavior of the eigenvalues λ_i as T tends to infinity. Specifically we derive (6.45), (6.58), (6.72), and (6.76).

1. Power Sums of Eigenvalues

Equations 6.45 and 6.58 involve a minor generalization of some known properties of Toeplitz forms [1]. The details are as follows.

It is known that the power sums of the eigenvalues λ_i of the complex correlation function can be expressed as [2]

$$\sum_i \lambda_i^k = \int R_k(t, t) \, dt, \tag{A4.1}$$

where

$$\tilde{R}_1(t, \tau) = R(t, \tau) \tag{A4.2a}$$

and

$$\tilde{R}_k(t, \tau) = \int \tilde{R}_1(t, x)\tilde{R}_{k-1}(x, \tau) \, dx \tag{A4.2b}$$

253

for $k > 1$. Moreover

$$R(t, \tau) = \int C(\alpha, \tau - t) \exp{-j\pi\alpha(t + \tau)} \, d\alpha, \qquad (A4.3a)$$

where

$$C(\alpha, \tau - t) = \mathcal{R}(\alpha, \tau - t)\theta(\tau - t, \alpha). \qquad (A4.3b)$$

Upon introducing (A4.3a) into (A4.2) and then into (A4.1), we obtain

$$\sum_i \lambda_i^k = \int \cdots \int C(x_1, y_1) \cdots C(x_{k-1}, y_{k-1})C(-x_1, \ldots, -x_{k-1}, -y_1, \ldots, -y_{k-1})$$

$$\times \exp j\pi \sum_{i=1}^{k-1} \left(\sum_{n=1}^{i-1} x_i y_n - \sum_{n=i+1}^{k-1} x_i y_n \right) d\mathbf{x} \, d\mathbf{y}. \qquad (A4.4)$$

where we have used $d\mathbf{x}$ and $d\mathbf{y}$ to denote $dx_1, dx_2 \cdots dx_k$ and $dy_1, dy_2 \cdots dy_k$.

We next express $C(\cdot, \cdot)$ in terms of the fixed waveform $u_o(t)$ from which $u(t)$ is derived. The appropriate expression is readily found to be

$$C(\alpha, \beta) = C_T(\alpha T, \beta), \qquad (A4.5a)$$

where

$$C_T(x, y) = \mathcal{R}\left(\frac{x}{T}, y\right)\theta_o\left(\frac{y}{T}, x\right) \qquad (A4.5b)$$

and

$$\theta_o(\tau, f) = \int u_o\left(t - \frac{\tau}{2}\right) u_o\left(t + \frac{\tau}{2}\right) \exp j2\pi ft \, dt. \qquad (A4.5c)$$

Upon introducing (A4.5) into (A4.4) and changing the variable of integration, we obtain

$$\frac{1}{T} \sum_i (T\lambda_i)^k = \int \cdots \int C_T(\alpha_1, \beta_1) \cdots C_T(\alpha_{k-1}, \beta_{k-1})$$

$$\times C_T(-\alpha_1 \cdots -\alpha_{k-1}, -\beta_1 \cdots -\beta_{k-1}) \exp j \frac{\pi}{T} \sum_{i=1}^{k-1} \alpha_i \left(\sum_{n=1}^{i-1} y_n - \sum_{n=i+1}^{k-1} y_n \right) d\boldsymbol{\alpha} \, d\boldsymbol{\beta}.$$

$$(A4.6)$$

We next let the duration T of $u(t)$ tend to infinity. For simplicity we restrict the two-frequency correlation function to be continuous—a rather mild constraint. This restriction, in conjunction with the continuity of $\theta_o(\tau, t)$ implies that the integrand of (A4.6) is continuous. Since the quantity $C_T(\alpha, \beta)$ tends to $\mathcal{R}(0, \beta)\theta_o(0, \alpha)$ as T tends to infinity, the integrand of (A4.6) converges to

$$\mathcal{R}(0, \beta_1) \cdots \mathcal{R}(0, \beta_{k-1})\mathcal{R}(0, -\beta_1 - \cdots - \beta_{k-1})\theta_o(0, \alpha_1) \cdots$$

$$\theta_o(0, \alpha_{k-1})\theta_o(0, -\alpha_1 - \cdots -\alpha_{k-1}).$$

We can also show that the integrand of (A4.6) is upper bounded by a function that is integrable over the region involved. Thus by the dominated convergence theorem, we obtain [3]

$$\lim_{T \to \infty} \frac{1}{T} \sum_i (T\lambda_i)^k = \left[\int \mathscr{R}(0, \beta_1) \cdots \mathscr{R}(0, \beta_{k-1}) \mathscr{R}(0, -\beta_1 - \cdots - \beta_{k-1}) \, d\beta \right]$$

$$\times \left[\int \theta_o(0, \alpha_1) \cdots \theta_o(0, \alpha_{k-1}) \theta(0, -\alpha_1 - \cdots - \alpha_{k-1}) \, d\alpha \right] \quad \text{(A4.7)}$$

Finally, we express $\mathscr{R}(0, \beta)$ and $\theta_o(0, \alpha)$ in terms of $\sigma(r, f)$ and $u_o(t)$ and then evaluate the integrals to obtain

$$\lim_{T \to \infty} \frac{1}{T} \sum_i (T\lambda_i)^k = \int [\sigma(f)]^k \, df \int |u_o(t)|^{2k} \, dt \quad \text{(A4.8)}$$

which is (6.58). Equation 6.45 is obtained from (A.48) by setting k equal to 2 and 3.

2. The Dominant Eigenvalue

To establish (6.72), we proceed as follows. It is known that the dominant eigenvalue λ_o satisfies the inequality

$$\lambda_o \le \left[\sum_i \lambda_i^{n+1} \right]^{1/n+1} \quad \text{(A4.9)}$$

for all positive integers n [4]. Consequently

$$\lim_{T \to \infty} \lambda_o T^{n/1+n} \le \lim_{T \to \infty} \left[\sum_i T^{-1} (\lambda_i T)^{n+1} \right]^{1/n+1}$$

or by virtue of (A4.8),

$$\lim_{T \to \infty} \lambda_o T^{n/1+n} \le \left\{ \int [\sigma(f)]^{n+1} \, df \int |u_o(t)|^{2n+2} \, dt \right\}^{1/n+1} \quad \text{(A4.10)}$$

We next over bound $[\sigma(f)]^{n+1}$ and $|u_o(t)|^{2n+2}$ by $\sigma_m [\sigma(f)]^n$ and $|u_m|^2 |u_o(t)|^{2n}$, respectively, where σ_m and $|u_m|$ are the maximum values of $\sigma(f)$ and $|u(t)|$. Introducing these bounds into (A4.10) yields

$$\lim_{T \to \infty} \lambda_o T^{n/1+n} \le [|u_m|^2 \sigma_m]^{n/n+1}$$

Finally we let n approach infinity to conclude that

$$\lim_{T \to \infty} \lambda_o T \le |u_m|^2 \sigma_m . \quad \text{(A4.11)}$$

To show that (A4.11) is in fact an equality, we employ the inequality

$$\lambda_o \geq \iint \chi(t) R(t, \tau) \chi^*(\tau) \, dt \, d\tau \tag{A4.12}$$

which is known to be valid for every function $\chi(t)$ of unit norm [5]. Specifically we take $\chi(t)$ to be of the form $(1/\sqrt{T}) \chi_o(t/T) \exp j2\pi \hat{f} t$ where $\hat{f}$ is specified subsequently and where $\chi_o(\cdot)$ does not depend upon T. We then combine (A4.3), (A4.5), and (A4.12) to obtain

$$\lambda_o \geq \frac{1}{T} \iiint C_T(\alpha T, \tau - t) \chi_o\left(\frac{t}{T}\right) \chi_o^*\left(\frac{\tau}{T}\right) \exp -j\pi[\alpha(t + \tau) - 2\hat{f}(t - \tau)] \, dt \, d\tau \, d\alpha$$

or upon changing the variables of integration,

$$T\lambda_o \geq \iiint C_T(y, \beta) \chi_o\left(x - \frac{\beta}{2T}\right) \chi_o^*\left(x + \frac{\beta}{2T}\right) \exp -j2\pi(xy + \hat{f}\beta) \, dx \, dy \, d\beta.$$

We next appeal to the dominated convergence argument that lead to (A4.7) to obtain

$$\lim_{T \to \infty} T\lambda_o \geq \iiint \mathscr{R}(0, \beta) \theta_o(0, y) |\chi_o(x)|^2 \exp -j2\pi(xy + \hat{f}\beta) \, dx \, dy \, d\beta,$$

or upon evaluating the integrals on y and β,

$$\lim_{T \to \infty} T\lambda_o \geq \sigma(\hat{f}) \int |u_o(t)|^2 |\chi_o(t)|^2 \, dt. \tag{A4.13}$$

It is clear that the right member of (A4.13) is maximized by choosing $\hat{f}$ so that $\sigma(\hat{f}) = \sigma_m$ and choosing $\chi_o(t)$ to be a constant at those points where $|u_o(t)| = |u_m|$ and zero elsewhere. With these choices we obtain

$$\lim_{T \to \infty} T\lambda_o \geq \sigma_m |u_m|^2 \int |\chi_o(t)|^2 \, dt = \sigma_m |u_m|^2. \tag{A4.14}$$

Since the right members of (A4.11) and (A4.13) are equal, we have established that

$$\lim_{T \to \infty} T\lambda_o = \sigma_m |u_m|^2 \tag{A4.15}$$

as was to be shown.

3. A Bound to the Dominant Eigenvalue

To demonstrate that the right member of (6.76) is an upper bound to the dominant eigenvalue λ_o, we proceed as follows. Let $\varphi_o(t)$ be the eigenfunction associated with λ_o; that is,

$$\lambda_o \varphi_o(t) = \int R(t, \tau) \varphi_o(\tau) \, d\tau, \tag{A4.16a}$$

or equivalently

$$\lambda_o \varphi_o(t) = \iiint \sigma(r, f) u(t - r) u^*(\tau - r) \varphi(\tau) \exp j2\pi f(\tau - t) \, dr \, df \, d\tau \tag{A4.16b}$$

Upon multiplying both members of the latter expression by $u^*(t - x) \exp 2\pi yt$ and integrating with respect to t, we obtain

$$\lambda_o G(x, y) = \iint \sigma(r, f)\theta(r - x, y - f)G(r, f) \exp j\pi(y - f)(x + r) \, dr \, df$$

(A4.17)

where we define

$$G(x, y) = \int \varphi_o(t)u^*(t - x) \exp j2\pi yt \, dt$$

(A4.18a)

and, as before,

$$\theta(\tau, f) = \int u\left(t - \frac{\tau}{2}\right)u^*\left(t + \frac{\tau}{2}\right) \exp j2\pi ft \, dt.$$

(A4.18b)

We next take magnitudes in (A4.17) and employ the inequality between the magnitude of an integral and the integral of the magnitude of the integrand to obtain

$$\lambda_o|G(x, y)| \leq \iint \sigma(r, f)|\theta(r - x, y - f)G(r, f)| \, dr \, df.$$

(A4.19)

Finally, we choose x and y so as to maximize the value of $|G(x, y)|$. Upon denoting these values by x_o and y_o, respectively, (A4.19) becomes

$$\lambda_o|G(x_o, y_o)| \leq \iint \sigma(r, f)|\theta(r - x_o, y_o - f)G(r, f)| \, dr \, df,$$

(A4.20)

or since $|G(r, f)|$ is no greater than $|G(x_o, y_o)|$,

$$\lambda_o \leq \iint \sigma(r, f)|\theta(r - x_o, y_o - f)| \, dr \, df$$

$$\leq \max_{f', r'} \iint \sigma(r, f)|\theta(r - r', f' - f)| \, dr \, df.$$

(A4.21)

B. HIGHLY DOUBLY DISPERSIVE CHANNELS

We now suppose that the complex envelope $u(t)$ is fixed and that the scattering function is of the form

$$\sigma(r, f) = \left(\frac{1}{a}\right)^2 \sigma_o\left(\frac{r}{a}, \frac{f}{a}\right),$$

where

$$\iint \sigma(r, f) \, dr \, df = \iint [\sigma_o(r, f)]^2 \, dr \, df = 1$$

so that the spread S of $\sigma(\cdot\,,\,\cdot)$ equals a^2. We seek the limiting values of

$$b = \sum_i \lambda_i^2$$

and

$$d = \sum_i \lambda_i^3$$

as a tends to infinity. To this end we note, from (A4.4), that

$$Sb = a^2 \int \cdots \int C(x_1, y_1)C(-x_1, -y_1)\, dx\, dy \qquad (A4.22a)$$

and

$$S^2 d = a^4 \int \cdots \int C(x_1, y_1)C(x_2, y_2)C(-x_1 - x_2, -y_1 - y_2)$$

$$\times \exp j\pi(x_2 y_1 - x_1 y_2)\, dx\, dy, \qquad (A4.22b)$$

where

$$C(x, y) = \mathcal{R}(x, y)\theta(y, x) \qquad (A4.22c)$$

and where $\theta(\cdot\,,\,\cdot)$ and $\mathcal{R}(\cdot\,,\,\cdot)$ are associated with $u(t)$ and $\sigma(r, f)$.

We next express $\mathcal{R}(\cdot\,,\,\cdot)$ in terms of $\mathcal{R}_o(\cdot\,,\,\cdot)$, the two-frequency correlation function associated with $\sigma_o(\cdot\,,\,\cdot)$, combine the results with (A4.22), and then change the variable of integration to obtain

$$Sb = \iint |\mathcal{R}_o(\alpha, \beta)|^2 \left| \theta\!\left(\frac{\beta}{a}, \frac{\alpha}{a}\right) \right|^2 d\alpha\, d\beta \qquad (A4.23a)$$

and

$$S^2 d = \int \cdots \int \mathcal{R}_o(\alpha_1, \beta_1)\mathcal{R}_o(\alpha_2, \beta_2)\mathcal{R}_o(-\alpha_1 - \alpha_2, -\beta_1 - \beta_2)$$

$$\times \theta\!\left(\frac{\beta_1}{a}, \frac{\alpha_1}{a}\right)\theta\!\left(\frac{\beta_2}{a}, \frac{\alpha_2}{a}\right)\theta\!\left(\frac{-\beta_1 - \beta_2}{a}, \frac{-\alpha_1 - \alpha_2}{a}\right)$$

$$\times \exp j\frac{\pi}{a}(\alpha_2\beta_1 - \beta_2\alpha_1)\, d\alpha\, d\beta. \qquad (A4.23b)$$

Finally, we repeat the dominated convergence argument involved in (A4.7) to conclude that

$$\lim_{S \to \infty} Sb = \iint |\mathcal{R}_o(\alpha, \beta)|^2\, d\alpha\, d\beta \qquad (A4.24a)$$

and

$$\lim_{S \to \infty} S^2 d = \int \cdots \int \mathcal{R}_o(\alpha_1, \beta_1)\mathcal{R}_o(\alpha_2, \beta_2)\mathcal{R}_o(-\alpha_1 - \alpha_2, -\beta_1 - \beta_2)\, d\alpha\, d\beta,$$

$$(A4.24b)$$

where we used the fact that $\theta(0, 0) = 1$. Upon expressing $\mathcal{R}_o(\cdot, \cdot)$ in terms of $\sigma_o(\cdot, \cdot)$, we obtain

$$\lim_{S \to \infty} Sb = \iint [\sigma_o(r, f)]^2 \, dr \, df = 1 \tag{A4.25a}$$

and

$$\lim_{S \to \infty} S^2 \, d = \iint [\sigma_o(r, f)]^3 \, dr \, df. \tag{A4.25b}$$

Introducing these expressions into (6.82) yields (6.85).

C. HIGH TW MODULATION

We now develop the relevant properties of $\bar{E}_u$, $\bar{b}_u$, and $\bar{d}_u$ over a Gaussian ensemble of waveforms $u(t)$. The development consists of three lemmas culminating in a theorem that summarizes the relevant results.

LEMMA 1. Let $u(t)$ be a sample function of a complex-valued zero-mean Gaussian random process possessing the correlation function $Q(t, \tau)$ and having statistically independent real and imaginary parts. The moments of this process satisfy the relation

$$\overline{u(t_1)u^*(\tau_1)u(t_2)u^*(\tau_2) \cdots u(t_n)u^*(\tau_n)} = \sum Q(t_1, \tau_{j_1})Q(t_2, \tau_{j_2}) \cdots Q(t_n, \tau_{j_n}),$$

where the summation is over all permutations $j_1, \ldots, j_n$ of the integers $1, \ldots, n$

Proof. This is a simple variation of a known result [6].

LEMMA 2. Let $u(t)$ be a sample function of a complex-valued zero-mean Gaussian random process with the correlation function $Q(t, \tau)$ and with statistically independent real and imaginary parts. In particular let

$$Q(t, \tau) = \frac{\sqrt{2}}{T} \exp -\pi \left[\left(\frac{t}{T} \right)^2 + \left(\frac{\tau}{T} \right)^2 + \frac{W^2}{2} (t - \tau)^2 \right]. \tag{A4.26}$$

Let $\theta_u(\alpha, \beta)$ be given by the expression

$$\theta_u(\alpha, \beta) = \int u \left(x - \frac{\alpha}{2} \right) u^* \left(x + \frac{\alpha}{2} \right) \exp j2\pi\beta x \, dx \tag{A4.27}$$

The moments of $\theta_u(\alpha, \beta)$ satisfy the following expressions:

$$\overline{\theta_u(0, 0)} = 1, \tag{A4.28}$$

$$\overline{|\theta_u(\alpha, \beta)|^2} = \{\exp - \pi[(\alpha W)^2 + (\beta T)^2]\} \exp - \pi\left(\frac{\alpha}{T}\right)^2$$

$$+ \frac{1}{TW\sqrt{1 + c}} \exp - \pi\left[(1 + c)^{-1}\left(\frac{\beta}{W}\right)^2 + \left(\frac{\alpha}{T}\right)^2\right], \tag{A4.29}$$

and

$$\overline{\theta_u(\alpha_1, \beta_1)\theta_u(\alpha_2, \beta_2)\theta_u(-\alpha_1 - \alpha_2, -\beta_1 - \beta_2)} = \sum_{i=1}^{6} A_i, \tag{A4.30}$$

where

$$A_1 = F_T \Phi(W\alpha_1)\Phi(W\alpha_2)\Phi(W\sqrt{\alpha_1\alpha_2})\Phi(T\beta_1)\Phi(T\beta_2)\Phi(T\sqrt{\beta_1\beta_2}),$$

$$A_2 = \frac{F_T}{TW(1 + c)} \Phi\left(\sqrt{\frac{3}{4}} W\alpha_1\right)\Phi\left(\sqrt{\frac{3}{4}} T\beta_1\right)\Phi\left(\frac{\beta_1 + 2\beta_2}{2W\sqrt{1 + c}}\right),$$

$$A_3 = \frac{F_T}{TW(1 + c)} \Phi\left(\sqrt{\frac{3}{4}} W\alpha_2\right)\Phi\left(\sqrt{\frac{3}{4}} T\beta_2\right)\Phi\left(\frac{\beta_2 + 2\beta_1}{2 W\sqrt{1 + c}}\right),$$

$$A_4 = \frac{F_T}{TW(1 + c)} \Phi\left(\sqrt{\frac{3}{4}} W(\alpha_1 + \alpha_2)\right)\Phi\left(\sqrt{\frac{3}{4}} T(\beta_1 + \beta_2)\right)\Phi\left(\frac{\beta_1 - \beta_2}{2W\sqrt{1 + c}}\right),$$

$$A_5 = \frac{4c}{(3 + 4c)} \Phi\left(\frac{2}{T}\left(\frac{1 + c}{3 + 4c}\right)^{1/2}\sqrt{\alpha_1{}^2 + \alpha_2{}^2 + \alpha_1\alpha_2}\right)$$

$$\times \Phi\left(\frac{2}{W}\left(\frac{\beta_1{}^2 + \beta_2{}^2 + \beta_1\beta_2}{3 + 4c}\right)^{1/2}\right) \exp - j\pi\left(\frac{\alpha_1\beta_2 - \alpha_2\beta_1}{1 + 4c/3}\right)$$

$$A_6 = A_5^*$$

with

$$F_T = \exp - \frac{\pi}{T^2}(\alpha_1{}^2 + \alpha_2{}^2 + \alpha_1\alpha_2),$$

$$\Phi(x) = \exp - \pi x^2,$$

$$c = (TW)^{-2}.$$

Proof. Equation A4.28 is proved by noting that

$$\overline{\theta_u(0, 0)} = \int |u(t)|^2 \, dt = \int \overline{|u(t)|^2} \, dt = \int Q(t, t) \, dt$$

and hence by virtue of (A4.26) that

$$\overline{\theta_u(0,0)} = \frac{\sqrt{2}}{T} \int \exp - 2\pi \left(\frac{t}{T}\right)^2 dt$$

which, upon evaluation of the integral, yields (A4.28).

To prove (A4.29) and (A4.30) we consider the average value of the quantity

$$\theta_u(\alpha_1, \beta_1)\theta_u(\alpha_2, \beta_2) = \Lambda_{12}.$$

This quantity is denoted by Λ_{12} for ease of notation.

By the definition of Λ_{12} and the properties of averages

$$\overline{\Lambda}_{12} = \iint \overline{u\left(x_1 - \frac{\alpha_1}{2}\right)u^*\left(x_1 + \frac{\alpha_1}{2}\right)u\left(x_2 - \frac{\alpha_2}{2}\right)u^*\left(x_2 + \frac{\alpha_2}{2}\right)}$$

$$\times \exp j2\pi(x_1\beta_1 + x_2\beta_2)\, dx_1\, dx_2, \quad \text{(A4.31)}$$

or by an application of Lemma 1

$$\overline{\Lambda}_{12} = \iint Q\left(x_1 - \frac{\alpha_1}{2}, x_1 + \frac{\alpha_1}{2}\right)Q\left(x_2 - \frac{\alpha_2}{2}, x_2 + \frac{\alpha_2}{2}\right)$$

$$+ Q\left(x_1 - \frac{\alpha_1}{2}, x_2 + \frac{\alpha_2}{2}\right)Q\left(x_2 - \frac{\alpha_2}{2}, x_1 + \frac{\alpha_1}{2}\right)$$

$$\times \exp j2\pi(x_1\beta_1 + x_2\beta_2)\, dx_1\, dx_2. \quad \text{(A4.32)}$$

We next introduce the expression of (A4.26) for $Q(\cdot, \cdot)$ into (A4.32) and collect terms to obtain

$$\overline{\Lambda}_{12} = \frac{2}{T^2} \iint \left\{ \exp - 2\pi \left[\left(\frac{x_1}{T}\right)^2 + \left(\frac{x_2}{T}\right)^2 + \left(\frac{\alpha_1}{2T}\right)^2 + \left(\frac{\alpha_2}{2T}\right)^2\right] \right\}$$

$$\times [\exp j2\pi(x_1\beta_1 + x_2\beta_2)]\left\{\exp - \frac{\pi W^2}{2}(\alpha_1^2 + \alpha_2^2)\right.$$

$$+ \left[\exp - \pi W^2(x_1 - x_2)^2 \exp - \frac{\pi W^2}{4}(\alpha_1 + \alpha_2)^2\right]\right\} dx_1\, dx_2. \quad \text{(A4.33)}$$

To proceed, we evaluate the integrals with respect to x_1 and x_2. This can be done most simply by casting the integral in the form of the matrix representation for two-dimensional Gaussian probability densities. Since the details of the evaluation are uninteresting, we merely state the result [7]. It is

$$\overline{\Lambda}_{12} = \left\{ \exp - \frac{\pi}{2} \left[(\alpha_1 W)^2 + (\alpha_2 W)^2 + (\beta_1 T)^2 + (\beta_2 T)^2 \right] \right\}$$

$$\times \exp - \frac{\pi}{2} \left[\left(\frac{\alpha_1}{T} \right)^2 + \left(\frac{\alpha_2}{T} \right)^2 \right] + \frac{1}{TW\sqrt{1+c}}$$

$$\times \left\{ \exp - \frac{\pi}{4} \left[W^2 (\alpha_1 + \alpha_2)^2 + T^2 (\beta_1 + \beta_2)^2 \right] \right\}$$

$$\times \exp - \frac{\pi}{2} \left[\frac{(\beta_1 - \beta_2)^2}{2(1+c)W^2} + \frac{\alpha_1{}^2 + \alpha_2{}^2}{T^2} \right], \qquad (A4.34a)$$

where we have introduced

$$c = \left(\frac{1}{TW} \right)^2. \qquad (A4.34b)$$

Equation A4.34 is used in the subsequent proof of (A4.30). It can also be specialized to prove the second statement of the lemma. Specifically since

$$|\theta_u(\alpha, \beta)|^2 = \theta_u(\alpha, \beta)\theta_u(-\alpha, -\beta),$$

the average value $\overline{|\theta_u(\alpha, \beta)|^2}$ can be evaluated from (A4.34) by setting α_1 equal to α, β_1 equal to β, α_2 equal to $-\alpha$, and β_2 equal to $-\beta$. Thus we obtain

$$\overline{|\theta_u(\alpha, \beta)|^2} = \{\exp - \pi[(\alpha W)^2 + (\beta T)^2]\} \exp - \pi \left(\frac{\alpha}{T} \right)^2$$

$$+ \frac{1}{TW\sqrt{1+c}} \exp - \pi \left[(1+c)^{-1} \left(\frac{\beta}{W} \right)^2 + \left(\frac{\alpha}{T} \right)^2 \right] \quad (A4.35)$$

which is just (A4.29).

We turn now to a proof of (A4.30). For ease of notation we define

$$\Lambda_{123} = \theta_u(\alpha_1, \beta_1)\theta_u(\alpha_2, \beta_2)\theta_u(\alpha_3, \beta_3), \qquad (A4.36a)$$

where

$$\alpha_3 = -(\alpha_1 + \alpha_2) \qquad (A4.36b)$$

and

$$\beta_3 = -(\beta_1 + \beta_2). \qquad (A4.36c)$$

By the definition of $\theta_u(\cdot, \cdot)$ and by the properties of averages, the average value of Λ_{123} can be expressed as

$$\overline{\Lambda_{123}} = \iiint \overline{u\left(x_1 - \frac{\alpha_1}{2}\right)u^*\left(x_1 + \frac{\alpha_1}{2}\right)u\left(x_2 - \frac{\alpha_2}{2}\right)u^*\left(x_2 + \frac{\alpha_2}{2}\right)u\left(x_3 - \frac{\alpha_3}{2}\right)}$$

$$\times \overline{u^*\left(x_3 + \frac{\alpha_3}{2}\right)} \exp j2\pi \sum_{i=1}^{3} x_i \beta_i \, dx_1 \, dx_2 \, dx_3 . \quad \text{(A4.37)}$$

We next appeal to Lemma 1 in order to expand the average of (A4.37) into a product of second-order averages. The resulting expansion is

$$\overline{\Lambda_{123}} = \sum_{i=1}^{6} A_i, \quad \text{(A4.38)}$$

where

$$A_1 = \prod_{i=1}^{3} \int Q\left(x_i - \frac{\alpha_i}{2}, x_i + \frac{\alpha_i}{2}\right) \exp j2\pi x_i \beta_i \, dx \ ,$$

$$A_2 = \left[\int Q\left(x_1 - \frac{\alpha_1}{2}, x_1 + \frac{\alpha_1}{2}\right) \exp j2\pi x_1 \beta_1 \, dx_1\right](\overline{\Lambda_{23}}) - A_1,$$

$$A_3 = \left[\int Q\left(x_2 - \frac{\alpha_2}{2}, x_2 + \frac{\alpha_2}{2}\right) \exp j2\pi x_2 \beta_2 \, dx_2\right](\overline{\Lambda_{13}}) - A_1,$$

$$A_4 = \left[\int Q\left(x_3 - \frac{\alpha_3}{2}, x_3 + \frac{\alpha_3}{2}\right) \exp j2\pi x_3 \beta_3 \, dx_3\right](\overline{\Lambda_{12}}) - A_1,$$

$$A_5 = \iiint Q\left(x_1 - \frac{\alpha_1}{2}, x_2 + \frac{\alpha_2}{2}\right)Q\left(x_2 - \frac{\alpha_2}{2}, x_3 + \frac{\alpha_3}{2}\right)Q\left(x_3 - \frac{\alpha_3}{2}, x_1 + \frac{\alpha_1}{2}\right)$$

$$\times \exp j2\pi \sum_{i=1}^{3} x_i \beta_i \, dx_1 \, dx_2 \, dx_3 ,$$

$$A_6 = A_5^*$$

and where Λ_{ij} is, except for a change of index, the function of (A4.31) or (A4.34), that is,

$$\overline{\Lambda_{ik}} = \left\{\exp - \frac{\pi}{2} [(\alpha_i W)^2 + (\alpha_k W)^2 + (\beta_i T)^2 + (\beta_k T)^2]\right\}$$

$$\times \exp - \frac{\pi}{2} \left[\left(\frac{\alpha_i}{T}\right)^2 + \left(\frac{\alpha_k}{T}\right)^2\right] + \frac{1}{TW\sqrt{1+c}}$$

$$\times \left\{\exp - \frac{\pi}{4} [W^2(\alpha_i + \alpha_k)^2 + T^2(\beta_i + \beta_k)^2]\right\}$$

$$\times \exp - \frac{\pi}{2} \left[\frac{(\beta_i - \beta_k)^2}{2(1 + c)W^2} + \frac{\alpha_i^2 + \alpha_k^2}{T^2}\right]. \quad \text{(A4.39)}$$

Our objective is to show that A_1 to A_5 are given by the expressions of the lemma.

The integrals appearing in the expressions for A_1 through A_4 can be evaluated readily for the correlation function $Q(\cdot, \cdot)$ of (A4.26). The results are

$$A_1 = \left\{ \exp - \frac{\pi}{2} \sum_{i=1}^{3} [(\alpha_i W)^2 + (\beta_i T)^2] \right\} \exp - \frac{\pi}{2} \sum_{i=1}^{3} \left(\frac{\alpha_i}{T} \right)^2, \tag{A4.40}$$

$$A_2 = \left\{ \exp - \frac{\pi}{2} [(\alpha_1 W)^2 + (\beta_1 T)^2] \right\} \left[\exp - \frac{\pi}{2} \left(\frac{\alpha_1}{T} \right)^2 \right] (\bar{\Lambda}_{23}) - A_1, \tag{A4.41a}$$

$$A_3 = \left\{ \exp - \frac{\pi}{2} [(\alpha_2 W)^2 + (\beta_2 T)^2] \right\} \left[\exp - \frac{\pi}{2} \left(\frac{\alpha_2}{T} \right)^2 \right] (\bar{\Lambda}_{13}) - A_1, \tag{A4.41b}$$

$$A_4 = \left\{ \exp - \frac{\pi}{2} [(\alpha_3 W)^2 + (\beta_3 T)^2] \right\} \left[\exp - \frac{\pi}{2} \left(\frac{\alpha_3}{T} \right)^2 \right] (\bar{\Lambda}_{12}) - A_1. \tag{A4.41c}$$

Upon introducing (A4.39) into (A4.41) combining the result with (A4.40), exploiting (A4.36), and introducing the functions $\Phi(\cdot)$ and F_r, we obtain the expressions of the lemma for A_1 through A_4.

Our remaining task is to evaluate A_5. To this end we introduce the expression of (A4.26), for $Q(\cdot, \cdot)$ into (A4.38), combine terms, eliminate α_3 and β_3 by use of (A4.36), and then, for conciseness, resort to matrix notation. As a result of these operations we obtain

$$A_5 = \frac{(2)^{3/2}}{T^3} \left\{ \exp - \pi \left[\left(\frac{W}{2} \right)^2 + T^{-2} \right] [\alpha_1{}^2 + \alpha_2{}^2 + \alpha_1 \alpha_2] \right\} [(2\pi)^{3/2} |\mathbf{D}|^{1/2}]$$

$$\times \left\{ \iiint \frac{(\exp - j2\pi \mathbf{y} \cdot \mathbf{x}') \exp - \frac{1}{2} \mathbf{x} \mathbf{D}^{-1} \mathbf{x}' \, d\mathbf{x}}{(2\pi)^{3/2} |\mathbf{D}|^{1/2}} \right\} \tag{A4.42}$$

where

$$\mathbf{x} = (x_1, x_2, x_3),$$

$$\mathbf{y} = (y_1, y_2, y_3)$$

with

$$y_1 = -\beta_1 + j \frac{W^2}{2} \left(\alpha_2 + \frac{\alpha_1}{2} \right), \tag{A4.43a}$$

$$y_2 = -\beta_2 - j \frac{W^2}{2} \left(\alpha_1 + \frac{\alpha_2}{2} \right), \tag{A4.43b}$$

$$y_3 = (\beta_1 + \beta_2) + j \frac{W^2}{4} (\alpha_1 - \alpha_2), \tag{A4.43c}$$

and

$$\mathbf{D}^{-1} = \pi W^2 \begin{bmatrix} 4c + 2 & -1 & -1 \\ -1 & 4c + 2 & -1 \\ -1 & -1 & 4c + 2 \end{bmatrix}. \qquad \text{(A4.44)}$$

In these expressions a superscript prime denotes transpose and $|\mathbf{D}|$ denotes the determinent of the matrix $\mathbf{D}$.

The integral of (A4.42) is recognizable as the characteristic function of the three-dimensional zero-mean Gaussian probability density with the co-variance matrix $\mathbf{D}$. Therefore the value of the integral is $\exp - 2\pi^2 \mathbf{y} \, \mathbf{D} \mathbf{y}'$ and hence

$$A_5 = \left(\frac{2\sqrt{\pi}}{T}\right)^3 |\mathbf{D}|^{1/2} \left\{ \exp - \pi \left[\left(\frac{W}{2}\right)^2 + T^{-2} \right] (\alpha_1^{\,2} + \alpha_2^{\,2} + \alpha_1 \alpha_2) \right\}$$

$$\times \exp - 2\pi^2 \mathbf{y} \mathbf{D} \mathbf{y}'. \quad \text{(A4.45)}$$

With the aid of (A4.43), we can verify that

$$\mathbf{y} \mathbf{D} \mathbf{y}' = \frac{1}{\pi(4c + 3)W^2} \left[2(\beta_1^{\,2} + \beta_2^{\,2} + \beta_1 \beta_2) - \frac{3}{8} W^4 (\alpha_1^{\,2} + \alpha_2^{\,2} + \alpha_1 \alpha_2) \right.$$

$$\left. + j \frac{3W^2}{2} (\alpha_1 \beta_2 - \beta_1 \alpha_2) \right]. \quad \text{(A4.46)}$$

Also

$$|\mathbf{D}|^{1/2} = \frac{TW}{2W^3 (\pi)^{3/2} (3 + 4c)}. \qquad \text{(A4.47)}$$

Upon combining (A4.45) through (A4.47), we obtain the expression of the lemma for A_5.

We are now ready to establish the limiting properties of $\bar{b}_u$, $\bar{d}_u$, and $\bar{E}_u$ over the ensemble of modulating waveforms.

LEMMA 3. Let $u(t)$ be a sample function of the random process described by the correlation function $Q(t, \tau)$ of Lemma 2, and let $R_u(t, \tau)$ be the complex correlation function that results from using $u(t)$ as the complex modulation in conjunction with a channel two-frequency correlation function $\mathscr{R}(\alpha, \beta)$. Let the eigenvalues of $R_u(t, \tau)$ be λ_i, $i = 1, \ldots$, and define

$$E_u = \int |u(t)|^2 \, dt,$$

$$b_u = \sum_i \lambda_i^{\,2},$$

$$d_u = \sum_i \lambda_i^{\,3}.$$

The average values $\bar{E}_u$, $\bar{b}_u$, and $\bar{d}_u$, of E_u, b_u, and d_u over the ensemble of sample functions $u(t)$ satisfy the following expressions:

$$\bar{E}_u = 1, \tag{A4.48a}$$

$$\lim_{\substack{T \to \infty \\ W \to \infty}} TW\bar{b}_u = 1 + S^{-1}, \tag{A4.48b}$$

$$\lim_{\substack{T \to \infty \\ W \to \infty}} (TW)^2 \, \bar{d}_u = \frac{4}{3}\left\{1 + h + 3S^{-1} + \iint [\sigma(r,f)]^3 \, dr \, df\right\}, \tag{A4.48c}$$

$$h = \int \cdots \int \sigma(r_1, f_1)\sigma(r_2, f_2)\sigma(r_3, f_3)$$
$$\times \exp -j2\pi[f_1(r_2 - r_3) + f_2(r_3 - r_1) + f_3(r_1 - r_2)]$$
$$\times dr_1 \, dr_2 \, dr_3 \, df_1 \, df_2 \, df_3. \tag{A4.48d}$$

Proof. It follows easily from (A4.27) that

$$E_u = \theta_u(0, 0)$$

and hence that

$$\bar{E}_u = \overline{\theta_u(0, 0)}$$

But according to (A4.28) of Lemma 2, $\overline{\theta_u(0, 0)}$ equals unity for the given ensemble. Thus $\bar{E}_u$ equals unity.

Let us now consider the average value $\bar{b}_u$ of b_u. Setting k equal to 2 in (A4.4) yields the familiar result

$$b_u = \iint |\mathscr{R}(\alpha, \beta)|^2 |\theta_u(\beta, \alpha)|^2 \, d\alpha \, d\beta. \tag{A4.49}$$

Or, upon taking the average value of b_u, interchanging the order of integrating and averaging, and introducing the expression of (A4.29) (Lemma 2) for $\overline{|\theta_u(\alpha, \beta)|^2}$, we obtain

$$\bar{b}_u = \iint |\mathscr{R}(\beta, \alpha)|^2 \Phi(\alpha W)\Phi(\beta T)\Phi\left(\frac{\alpha}{T}\right) d\alpha \, d\beta$$
$$+ \frac{1}{(TW)} \iint \frac{|\mathscr{R}(\beta, \alpha)|^2}{\sqrt{1 + c}} \Phi\left(\frac{1}{\sqrt{1 + c}}\left(\frac{\beta}{W}\right)\right)\Phi\left(\frac{\alpha}{T}\right) d\alpha \, d\beta, \tag{A4.50a}$$

where, as before

$$\Phi(x) = \exp - \pi x^2, \tag{A4.50b}$$

$$c = \left(\frac{1}{TW}\right)^2. \tag{A4.50c}$$

The magnitude of the second integral in the equation is bounded from above by $|\mathcal{R}(\beta, \alpha)|^2$, which we take to be an integrable function. Moreover as T and W tend to infinity and hence c tends to zero, the integrand converges to $|\mathcal{R}(\beta, \alpha)|^2$. Therefore by a dominated convergence argument [3], the corresponding integral converges to

$$\iint |\mathcal{R}(\beta, \alpha)|^2 \, d\alpha \, d\beta,$$

which is just the reciprocal of S, the channel spread factor. Consequently

$$\lim_{\substack{T \to \infty \\ W \to \infty}} TW \bar{b}_u = S^{-1} + \lim_{\substack{T \to \infty \\ W \to \infty}} TW \iint |\mathcal{R}(\beta, \alpha)|^2 \Phi(\alpha W)\Phi(\beta T)\Phi\left(\frac{\alpha}{T}\right) d\alpha \, d\beta.$$

$$(A4.51)$$

It remains to show that the rightmost limit equals unity. To this end, we first replace αW with a' and βT with β' to obtain the following expression for the limit:

$$\lim_{\substack{T \to \infty \\ W \to \infty}} \iint \left| \mathcal{R}\left(\frac{\beta'}{T}, \frac{\alpha'}{W}\right) \right|^2 \Phi(\alpha')\Phi(\beta')\Phi\left(\frac{\alpha'}{TW}\right) d\alpha' \, d\beta'$$

The integrand appearing in this expression is dominated by $\Phi(\alpha')\Phi(\beta')$, which is integrable, and converges to $\Phi(\alpha')\Phi(\beta')$ with increasing T and W. Consequently the dominated convergence theorem implies that the limit in question is

$$\iint \Phi(\alpha')\Phi(\beta') \, d\alpha' \, d\beta',$$

which is in fact equal to unity, as was to be proved.

Our final task is to prove (A4.48c). By virtue of (A4.3b) and (A4.4) with k set equal to 3,

$$\bar{d}_u = \int \cdots \int \mathcal{R}(\alpha_1, \beta_1)\mathcal{R}(\alpha_2, \beta_2)\mathcal{R}(-\alpha_1 - \alpha_2, -\beta_1 - \beta_2)$$

$$\times [\exp j\pi(\alpha_2\beta_1 - \alpha_1\beta_2)]\overline{\theta(\alpha_1, \beta_1)\theta(\alpha_2, \beta_2)\theta(-\alpha_1 - \alpha_2, -\beta_1 - \beta_2)}$$

$$\times d\alpha_1 \, d\alpha_2 \, d\beta_1 \, d\beta_2. \qquad (A4.52)$$

To proceed, we appeal to Lemma 2. That lemma permits us to express the average appearing in (A4.52) as

$$\sum_{i=1}^{6} A_i,$$

where the A_i are as defined in Lemma 2. Consequently $\bar{d}_u$ may be expressed as

$$\bar{d}_u = \left(\frac{1}{TW}\right)^2 \sum_{i=1}^{6} B_i, \tag{A4.53}$$

where

$$B_i = (TW)^2 \int \cdots \int \mathscr{R}(\alpha_1, \beta_1)\mathscr{R}(\alpha_2, \beta_2)\mathscr{R}(-\alpha_1 - \alpha_2, -\beta_1 - \beta_2)$$
$$\times [\exp j\pi(\alpha_2\beta_1 - \alpha_1\beta_2)]A_i \, d\alpha_1 \, d\alpha_2 \, d\beta_1 \, d\beta_2. \tag{A4.54}$$

We now evaluate each of the limits

$$\lim_{\substack{T \to \infty \\ W \to \infty}} B_i, \qquad i = 1, \ldots, 6.$$

By virtue of (A4.54) and Lemma 2, the first such limit can be cast in the following form (the variables of integration are changed from α_1 and α_2 to α_1/W and α_2/W and from β_1 and β_2 to β_1/T and β_2/T):

$$\lim_{\substack{T \to \infty \\ W \to \infty}} B_1 = \int \cdots \int \mathscr{R}\left(\frac{\alpha_1}{W}, \frac{\beta_1}{T}\right)\mathscr{R}\left(\frac{\alpha_2}{W}, \frac{\beta_2}{T}\right)\mathscr{R}\left(-\frac{\alpha_1 + \alpha_2}{W}, \frac{\beta_1 + \beta_2}{T}\right)$$

$$\times \left[\exp j\pi\left(\frac{\alpha_2\beta_1 - \alpha_1\beta_2}{TW}\right)\right][\exp -\pi(\alpha_1{}^2 + \alpha_2{}^2 + \alpha_1\alpha_2 + \beta_1{}^2 + \beta_2{}^2 + \beta_1\beta_2)]$$

$$\times \left[\exp -\frac{\pi}{(TW)^2}(\alpha_1{}^2 + \alpha_2{}^2 + \alpha_1\alpha_2)\right] d\alpha_1 \, d\alpha_2 \, d\beta_1 \, d\beta_2. \tag{A4.55}$$

The integrand of (A4.55) is dominated by the integrable function

$$\exp -\pi(\alpha_1{}^2 + \alpha_2{}^2 + a_1\alpha_2 + \beta_1{}^2 + \beta_2{}^2 + \beta_1\beta_2).$$

Moreover as T and W tend to infinity, the integrand converges to this function. Therefore by a dominated convergence argument,

$$\lim_{\substack{T \to \infty \\ W \to \infty}} B_1 = \int \cdots \int \exp -\pi(\alpha_1{}^2 + \alpha_2{}^2 + \alpha_1\alpha_2 + \beta_1{}^2 + \beta_2{}^2 + \beta_1\beta_2)$$

$$\times d\alpha_1 \, d\alpha_2 \, d\beta_1 \, d\beta_2$$

or

$$\lim_{\substack{T \to \infty \\ W \to \infty}} B_1 = \frac{4}{3}. \tag{A4.56}$$

The integral for B_2 is obtained from Lemma 2 and (A4.54). Replacing the variables of integration α_1 and β_1 by α_1/W and β_1/T yields an integrand that converges to the function

$$\mathscr{R}(0, 0)|\mathscr{R}(\alpha_2, \beta_2)|^2 \exp -\tfrac{3}{4}\pi[(\alpha_1)^2 + (\beta_1)^2]$$

and that is also bounded by an integrable function for all values of T and W. Hence

$$\lim_{\substack{T \to \infty \\ W \to \infty}} B_2 = \left[\iint |\mathscr{R}(\alpha, \beta)|^2 \, d\alpha \, d\beta \right] \left[\exp - 3\pi \left(\frac{\alpha}{2} \right)^2 \, d\alpha \right]^2$$

or equivalently

$$\lim_{\substack{T \to \infty \\ W \to \infty}} B_2 = \frac{4}{3} S^{-1}.$$

The derivations of the limiting values of B_3 and B_4 differ only trivially from that of B_2. The results are

$$\lim_{\substack{T \to \infty \\ W \to \infty}} B_i = \frac{4}{3} S^{-1}, \qquad i = 2, 3, 4. \tag{A4.57}$$

To obtain the limit of B_6 we combine (A4.54) with the result of Lemma 2 and then observe that the resulting integrand converges to

$$\frac{4}{3} \mathscr{R}(\alpha_1, \beta_1) \mathscr{R}(\alpha_2, \beta_2) \mathscr{R}(-\alpha_1 - \alpha_2, -\beta_1 - \beta_2)|.$$

and is dominated by

$$\frac{4}{3} |\mathscr{R}(\alpha_1, \beta_1) \mathscr{R}(\alpha_2, \beta_2) \mathscr{R}(-\alpha_1 - \alpha_2, -\beta_1 - \beta_2)|.$$

It is easy to show that the dominating function is integrable if $|\mathscr{R}(\alpha, \beta)|$ is integrable. Since in some problems of interest $|\mathscr{R}(\alpha, \beta)|$ is not integrable, we prefer to require only that $|\mathscr{R}(\alpha, \beta)|^2$ be integrable, that is, that S be nonzero. This supposition ensures that the dominating function is L_2, and we can invoke a strengthened form of the dominated convergence theorem to obtain [8]

$$\lim B_6 = \frac{4}{3} \int \cdots \int \mathscr{R}(\alpha_1, \beta_1) \mathscr{R}(\alpha_2, \beta_2)$$
$$\times \mathscr{R}(-\alpha_1 - \alpha_2, -\beta_1 - \beta_2) \, d\alpha_1 \, d\alpha_2 \, d\beta_1 \, d\beta_2.$$

This can be expressed more simply as

$$\lim_{\substack{T \to \infty \\ W \to \infty}} B_6 = \frac{4}{3} \iint [\sigma(r, f)]^3 \, dr \, df. \tag{A4.58}$$

Finally, an almost identical analysis of B_5 yields

$$\lim_{\substack{T \to \infty \\ W \to \infty}} B_5 = \frac{4}{3} \int \cdots \int \mathcal{R}(\alpha_1, \beta_1)\mathcal{R}(\alpha_2, \beta_2)\mathcal{R}(-\alpha_1 - \alpha_2, -\beta_1 - \beta_2)$$
$$\times \exp j2\pi(\alpha_2\beta_1 - \alpha_1\beta_2) \, d\alpha_1 \, d\alpha_2 \, d\beta_1 \, d\beta_2,$$

or equivalently

$$\lim_{\substack{T \to \infty \\ W \to \infty}} B_5 = \frac{4}{3} \int \cdots \int \sigma(r_1, f_1)\sigma(r_2, f_2)\sigma(r_3, f_3)$$
$$\times \exp -j2\pi[f_1(r_2 - r_3) + f_2(r_3 - r_1) + f_3(r_1 - r_2)]$$
$$\times dr_1 \, dr_2 \, dr_3 \, df_1 \, df_2 \, df_3. \tag{A4.59}$$

To complete the proof, we note that, by virtue of (A4.53),

$$\lim_{\substack{T \to \infty \\ W \to \infty}} (TW)^2 \, \bar{d}_u = \lim_{\substack{T \to \infty \\ W \to \infty}} \sum_{i=1}^{6} B_i = \sum_{i=1}^{6} \lim_{\substack{T \to \infty \\ W \to \infty}} B_i. \tag{A4.60}$$

The statement of the lemma follows directly from (A4.56) through (A4.60).

We are prepared to establish the results employed in Chapter 6. These results follow easily from Lemma 3.

THEOREM 1. Let $u(t)$ be a sample function of the complex-valued zero-mean Gaussian random process with statistically independent real and imaginary parts described by the correlation function $Q(t, \tau)$ with

$$Q(t, \tau) = \frac{\sqrt{2}}{T} \exp - \pi\left[\left(\frac{t}{T}\right)^2 + \left(\frac{\tau}{T}\right)^2 + \frac{W^2}{2}(t - \tau)^2\right].$$

Let $\bar{E}_u$, $\bar{b}_u$, and $\bar{d}_u$ be as defined in Chapter 6 and Lemma 3. Then

$$\bar{E}_u = 1, \tag{A4.61a}$$

$$\lim_{\substack{T \to \infty \\ W \to \infty}} TW\bar{b}_u = 1 + S^{-1}, \tag{A4.61b}$$

and

$$\lim_{\substack{T \to \infty \\ W \to \infty}} (TW)^2 \, \bar{d}_u = \frac{4}{3}\left\{1 + h + 3S^{-1} + \iint [\sigma(r, f)]^3 \, dr \, df\right\}, \tag{A4.61c}$$

where S is the channel spread factor and where

$$h = \int \cdots \int \sigma(r_1, f_1)\sigma(r_2, f_2)\sigma(r_3, f_3)$$
$$\times \exp -j2\pi[f_1(r_2 - r_3) + f_2(r_3 - r_1) + f_3(r_1 - r_2)]$$
$$\times dr_1 \, dr_2 \, dr_3 \, df_1 \, df_2 \, df_3. \tag{A4.61d}$$

Also

$$\lim_{\substack{T \to \infty \\ W \to \infty}} (TW)^2 \, \bar{d}_u \le \frac{4}{3} \left\{ 2 + 3S^{-1} + \iint [\sigma(r,f)]^3 \, dr \, df \right\} \qquad \text{(A4.62)}$$

and

$$\lim_{\substack{T \to \infty \\ W \to \infty}} (TW)^2 \, \bar{d}_u \le \frac{4}{3} [2 + (3 + \sigma)S^{-1}], \qquad \text{(A4.63)}$$

where

$$\sigma = \max_{r, f} \sigma(r, f).$$

Proof. Equation A4.61 is just a restatement of the result of Lemma 3. Equation A4.62 follows from (A4.61c) and (A4.61d) upon overestimating h by the integral of the magnitude of the integrand. Equation (A4.63) follows from (A4.62) upon noting that

$$\iint [\sigma(r,f)]^3 \, dr \, df \le \sigma \iint [\sigma(r,f)]^2 \, dr \, df = \sigma S^{-1}.$$

REFERENCES

[1] U. Grenander and G. Szego, *Toeplitz Forms and Their Applications*. Berkeley and Los Angeles: University of California Press, 1958.

[2] F. Smithies, *Integral Equations*. London: Cambridge University Press, 1958, Chapter 7.

[3] W. Rudin, *Principles of Mathematical Analysis*. New York: McGraw-Hill, 1953, pp. 209–210.

[4] F. G. Tricomi, *Integral Equations*. New York: Interscience, 1957, p. 124.

[5] F. Smithies, *Integral Equations*. London: Cambridge University Press, 1958, p. 131.

[6] J. M. Wozencraft and I. M. Jacobs, *Principles of Communication Engineering*. New York: Wiley, 1965, p. 205.

[7] Ibid., p. 52.

[8] E. J. McShane, *Integration*. Princeton: Princeton University Press, 1944, p. 174.

Glossary of Notation

$\rho_i{}^2$ Energy scattering cross section of ith scatterer (p. 10).

E_r Average received energy (p. 15).

$\sigma(r, f)$ Channel scattering function (p. 17).

$\sigma(r)$ Delay scattering function (p. 39).

$\sigma(f)$ Frequency scattering function (p. 39).

$s(t)$ Channel input waveform (p. 17).

$S(f)$ Fourier transform of $s(t)$ (p. 46).

$u(t)$ Complex envelope of $s(t)$ (p. 17).

$u_i(t)$ Complex envelope of the ith modulator waveform (p. 70).

$U(f)$ Fourier transform of $u(t)$ (p. 17).

$y(t)$ Channel output process (p. 17).

y_i Karhunen-Loève coefficient of $y(t)$ (p. 22).

$Y(f)$ Fourier transform of $y(t)$ (p. 46).

$z(t)$ Normalized complex envelope of $y(t)$ (p. 17).

z_i Karhunen-Loève coefficient of $z(t)$ (p. 25).

$Z(f)$ Fourier transform of $z(t)$ (p. 52).

$R(t, \tau)$ Complex correlation function (p. 17).

$R_k(t, \tau)$ Complex correlation function associated with the kth modulator waveform (p. 73).

$\tilde{R}_i(t, \tau)$ ith iterate of $R(t, \tau)$ (p. 186).

$R_y(t, \tau)$ Correlation function of $y(t)$ (p. 17).

$\varphi_i(t)$ Eigenfunction of $R(t, \tau)$ (p. 24).

$\varphi_{ik}(t)$ ith eigenfunction of $R_k(t, \tau)$ (p. 73).

λ_i Eigenvalue of $R(t, \tau)$ and fractional path strength in diversity interpretation (p. 24).

λ_0	Dominant eigenvalue; the largest of the eigenvalues λ_i, $i = 1, \ldots$ (p. 115).
λ	Any upper bound to the eigenvalues λ_i (p. 172).
$\mathscr{R}(\alpha, \beta)$	Two-frequency correlation function (p. 18).
$\theta(\tau, f)$	Two-dimensional correlation function of $u(t)$ (p. 19).
ω_0	Carrier frequency, radians per second (p. 17).
ω_i	Doppler shift of the ith scatterer, radians per second (p. 12).
$\tilde{\omega}_k$	Frequency of kth waveform in frequency position modulation (p. 78).
Δ	Frequency offset (Hz) between adjacent waveforms in frequency position modulation (p. 73).
B	Channel Doppler spread, frequency dispersion (p. 39).
L	Channel multipath spread, time dispersion (p. 39).
S	Total channel spread (p. 39).
T	Waveform duration (p. 40).
W	Waveform bandwidth (p. 40).
T_c	System coherence time (p. 53).
W_c	System coherence bandwidth (p. 53).
D	Number of effective diversity paths, or eigenvalues, in a system (p. 56).
D_i	Amount of diversity implicit in basic modulator waveform (p. 155).
D_e	Amount of explicit diversity employed (p. 155).
D^o	Number of effective diversity paths, or eigenvalues, in an optimized system. (p. 121).
b	$\sum_i \lambda_i^2$ (p. 62).
d	$\sum_i \lambda_i^3$ (p. 143).
K	Number of information bits per code word (p. 68).
m	Number of modulator waveforms (p. 68).
ν	$\mathrm{Log}_2\, m$.
$N_0/2$	Bilateral noise power density W/Hz (p. 68).
R	Information rate, bits per sec (p. 68).
r	Channel signaling rate, waveforms per sec (p. 68).
$r(t)$	Received waveform (p. 68).
r_{in}	(Normalized) Karhunen-Loève coefficient of $r(t)$ for one shot transmission (p. 78).

r_{inw} (Normalized) Karhunen-Loève coefficient of $r(t)$ for transmission of sequences (p. 83).

α Average received energy-to-noise power density ratio E_r/N_0 (p. 80).

α_p Average received energy-to-noise ratio per diversity path in an equal-strength diversity system (p. 136).

$\alpha_p{}^\circ$ Average received energy-to-noise ratio per diversity path in an optimized system (p. 120).

β Average received energy-to-noise ratio per information bit, α/ν (p. 108).

P Average power received during the basic transmission interval, τ (p. 105).

C Capacity of an infinite bandwidth additive white Gaussian noise channel $P/N_0 \ln 2$ (p. 105).

τ $1/r$; time allotted to the transmission of a modulator waveform (p, 105).

f_k Output of the kth branch demodulator in a one shot system (p. 82).

f_{kw} Output of the kth branch demodulator in response to the wth transmission in a sequence (p. 84).

$g_0(s)$ Moment-generating function of the "correct" branch demodulator output (p. 88).

$g_1(s)$ Moment-generating function of incorrect branch demodulator outputs' (p. 88).

$P(\varepsilon)$ System error probability (p. 90).

E An exponent in the bounds to $P(\varepsilon)$ for orthogonal waveforms (pp. 106, 109, 111).

E° The maximum value of E (p. 117).

E_b An exponent in the bounds to $P(\varepsilon)$ for orthogonal waveforms. Reliability per information bit, base 2 (pp. 108, 113).

$E_b{}^\circ$ The maximum value of E_b (p. 126).

$E(\rho)$ A function appearing in the expressions for E and E_b (p. 112).

$f_\rho(x)$ Eq. 5.27. (p. 118)

K_1, K_2 Coefficients in the bounds to $P(\varepsilon)$ for orthogonal waveforms (pp. 106, 107, 109, 114).

$\overline{P(\varepsilon)}$ System error probability averaged over an ensemble of codes (p. 91).

N Number of modulator waveforms per code word (p. 91). Also the amount of explicit diversity employed (p. 93).

E_c An exponent in the random coding bound (p. 91).

μ_o	A measure of the efficiency of a low-rate equal-strength diversity system relative to the optimized system (p. 139).
ε	A measure of system efficiency relative to an equal-strength diversity system (pp. 141, 153).
ε_∞	Value of ε for highly frequency dispersive mode of operation with D_i set to b^3/d^2 (p. 163).
ε'_∞	Value of ε for highly frequency dispersive mode of operation with D_i set to b/λ_o (p. 172).
ε_t	True efficiency of low-rate system relative to an optimized equal-strength system (p. 169).
$\tilde{\varepsilon}$	Measure of efficiency of filter-squarer system relative to nondispersive system (p. 195).
$\varepsilon_1, \varepsilon_2$	Measure of efficiency of a system employing a correlation kernel demodulator relative to equal-strength systems (pp. 191, 192).
$h(t, \tau)$	Demodulator kernel (p. 185).
Re	Real part of . . .
ln	Natural logarithm.
$\delta(\cdot)$	The unit impulse function
$\approx$	Approximately equal (p. 116).
$\sim$	Asymptotically equal (p. 107).
🦁	Optional section (p. 19).

Author Index

277

Subject Index

Ambiguity functions, of chirped pulse, 63, 158
  definition, 19
  idealized thumbtack, 65
Average power, 51, 105
Average received energy, 15, 78, 106

Bandwidth, coherence, 55
  definition, 41
  transmission, 3, 5, 95
Basic modulator waveform, 72-74
Branch demodulators, 1, 3, 85-89, 185-187

Capacity, of additive noise channel, 106, 109
  of fading channel, 110, 111, 125, 170, 171
Carrier frequency, 9, 185
Channel, dispersive only in frequency, 46-49, 162-175, 199
  dispersive only in time, 43-46, 175, 176, 199-204
  doubly dispersive, 49-51, 176-181
  nondispersive, 42, 156, 193, 194
  overspread, 148, 165, 177, 181
  singly dispersive, 43-49, 161, 191-204
  slightly dispersive, 156-161, 193-195
  underspread, 148
  wide sense stationary uncorrelated scattering, 18, 21

Chirped Gaussian pulse, 55, 63, 158, 160, 196
Coded systems, 69-71, 91, 130-132
  as diversity systems, 133
  efficiency of, 133-135
  random coding, 91-93
Coefficient, asymptotic, 107, 114
  for equal eigenvalues, 115
  lower bound, 107, 114, 115
  upper bound, 106, 114, 115
Coherence bandwidth, 35, 55
Coherence time, 35, 53-55
Complex correlation function, definition, 16
  eigenfunctions of, 22, 74
  eigenvalues of, 22, 73
  of received process, 13-29, 73, 74, 78
Complex envelopes, definition, 9
  filtered, 194
  of modulator waveform, 73
  of received waveform, 16, 25, 42, 43, 49, 50
Correlation bandwidth, 35, 53
Correlation function, definition, 13
  eigenfunctions, 24, 25
  eigenvalues, 24, 25
  of received process, 9, 13-29
Correlation kernel demodulators, 187-193
Correlation time, 53-55

279